Creation, Environment and Ethics

Creation, Environment and Ethics

Edited by

Rebekah Humphreys and Sophie Vlacos

CAMBRIDGE SCHOLARS
PUBLISHING

Creation, Environment and Ethics,
Edited by Rebekah Humphreys and Sophie Vlacos

This book first published 2010

Cambridge Scholars Publishing

12 Back Chapman Street, Newcastle upon Tyne, NE6 2XX, UK

British Library Cataloguing in Publication Data
A catalogue record for this book is available from the British Library

ISBN (10): 1-4438-2508-5, ISBN (13): 978-1-4438-2508-5

TABLE OF CONTENTS

Foreword

Mary Midgley

How many philosophers are there who can manage to write all the time about something interesting? Leafing through the list of Robin Attfield's works, my eye is continually caught by topics that I want at once to hear more about—wildernesses, animals, euthanasia, Thomas Traherne, unfathomed caves, secular humanism, biocentrism, efficiency in universities, the Continental Shelf, the Precautionary Principle, International Justice and the Third World, Aldo Leopold and Humpty Dumpty. All this on top of the Good of Trees, which is one of those discussions that one never forgets about, so that every oak and beech must surely still be thanking him for bringing the matter to the public attention.

It is really remarkable how Attfield has resisted the deadly tribalism of our age which shuts academic subjects up in separate cells in case they might infect one another. And his feat in doing this is especially surprising since that quarantine is imposed with special zeal over religion, which is thought to be particularly infectious. Long ago, when he (and I) first started writing, philosophers used to police these frontiers pretty sternly. The reproach of being "not real philosophy" hung over every topic that had not been mentioned during the last year in *Mind.* Philosophers did indeed know that some partly civilized territories existed on the borders of philosophy proper, notably philosophy of science and philosophy of religion. But prudent graduate-students were warned to avoid these places and their inhabitants were not supposed to intrude into serious, central business. Certainly nobody was expected to let loose, not just one of them but two simultaneously by using a title like *Creation, Evolution and Meaning*. This sort of combination might not have surprised Spinoza, Aristotle, Kant or indeed most of the mighty dead who are still named in our current syllabuses. But it was held that the times had become sterner and such laxness could no longer be tolerated.

These regulations, however, never deterred Attfield. By flatly ignoring the vetoes he has played a great part in bringing down some of the walls, so that travel between topics has become a good deal easier today. Indeed, with the mixed courses that are now so common, "philosophy of" this, that and the other is now quite widespread, and even sometimes gets departments and courses of its own. But of course, since academic tribalism is endemic, this can be a mixed blessing, leading to the building of new walls enclosing new enclaves… In fact, it is certainly not time for Attfield to stop writing. His spirit is going to be needed for a long time yet.

INTRODUCTION

REBEKAH HUMPHREYS
AND
SOPHIE VLACOS

Environmental ethics rose to prominence in the socially committed atmosphere of the early 1970s, an era beset by social and political uncertainties quite different from our own, and yet an era which has proved instrumental to the ongoing tenor of much of our social, political and ethical debate today. Since the time of its inception, when activists began to demand a philosophy of the environment, environmental ethics has naturally laid its emphasis upon practice, upon the need for practical solutions to the very real and present ecological dangers we now know so well; pollution, the depletion of our natural resources, our treatment of nonhuman species and overpopulation to name but a few. At the same time however, the challenges posed by environmental damage and the call to positive action have instigated a number of challenges at the theoretical level, compelling the renegotiation of a number of longstanding meta-ethical and indeed broader epistemological attitudes, towards rights and entitlements, the nature of humankind's moral responsibilities to the nonhuman world for example, and towards the philosopher's relation to empirical research, the bearing of scientific enquiry, and the very nature of ethical discourse. To this extent environmental ethics has proved itself a resolutely open and mixed discourse, one which has welcomed the challenges of global politics and scientific research as much as it has the general moral support of the wider intellectual community. The temptation of course, is to suppose that it could never have been otherwise, since environmental ethics must by its very nature be a reflective discipline, to an extent dependent upon the economic and ecological scrutiny of hard science and the shifting allegiances of the international community. Still, the blueprint for environmental ethics today cannot not be associated with the spirit of the age in which it first came to prominence, or with the overriding commitments of its founding voices. It is surely this overwhelming consensus which has enabled otherwise partisan representatives, theists and atheists for example, to participate in fruitful exchanges—regarding the compatibility of evolution and creation, or the compossibility of a beneficent God, of natural suffering and human wrongdoing—that are of equal principled import to all philosophers, regardless of one's environmental or ethical credentials?

A uniquely engaged, responsive and diverse branch of philosophy then, environmental ethics has developed whilst maintaining the usefulness and the purposefulness of its distinction within wider ethical discourse. To this extent Robin Attfield, a stalwart of the environmental-ethical community from the very beginning and a guiding voice in the development of the field, can be seen as something of an exemplary figure in this most exemplary of philosophical branches. One imagines that he would characterise himself as an environmental ethicist first and foremost, but he is a specialist of plural interests, with a distinguished career in ethics punctuated by some first-rate aesthetic, linguistic and theological reflections as well. For the non specialists among us, one of the most striking elements to emerge from Attfield's career and our contributors' reflections upon his work is the prevailing sense of restrained and rigorously garnered optimism underlying his intellectual journey and the philosophical positions he defends. It is with this sense of optimism and, if we may hazard, wonder for the natural world, that Attfield's environmental, aesthetic and theological interests are combined. In the first chapter Attfield writes in some detail about his career and the paths which led him to the ethical positions he adopts. Naturally, there is no competing with this personal testament and so we leave it to Attfield to elaborate upon this journey.

Given the tenor and indeed the duration of Attfield's career it was not hard to find an appropriately varied, willing and furthermore impressive list of contributors for this book. Wishing to celebrate Attfield's career and most importantly, to provide an accurate representation of his impact, we think it of particular advantage to be able to present our contributors' work with accompanying responses from Attfield. These comments follow individually as separate chapters between the works of our contributors. Once more we leave Attfield to present his thoughts and arguments on these matters. What follows therefore is a brief précis of our contributors' arguments.

In Chapter Two Nigel Dower tackles Attfield's ethics of climate change, emphasising the latter's hallmark commitments to the ethical considerations of both future generations and non-human species. Attfield's biocentrism and consequentialism do not however undermine his overwhelming conviction that climate change and the discourse of environmental ethics must be treated in conjunction with questions of human suffering and poverty in the developing world. For Attfield, and indeed for Dower, human development and environmental action are complimentary discourses; the alleviation of poverty and the alleviation of an increasingly choked and depleted planet are the aspirations of a common ethic with shared aims and goals. Such a desideratum leads both Attfield and Dower to reject the possibility of a single-handed approach to climate change (Dower, 19).

In the first part of his paper Dower examines the unilateral proposals for climate change represented by the Kyoto Protocol of 1990 (whereby a country's historical emissions were calculated upon the basis of their output in 1990 and a projective forecast for reductions made upon that basis), and by the aggregate emission proposal (wherein industrialised nations are penalised relative to their historical culpability). Dower explains why Attfield largely rejects the principle of historical indemnity as a means to calibrating future entitlement, aligning it to Attfield's wider preference for non-historical ethical principles. On the pragmatic level, historical approaches would inevitably lead to disagreement amongst countries as to their relative measures of past emissions. More generally, past emissions would set the benchmark for future goals irrespective of future demands and a changing global demographic, presently evident in the rapid industrialisation of nations such as China and India.

Whilst Attfield accepts the general principle of the Atmospheric Commons and the need to apportion these commons relative to national needs, and whilst in principle he accepts the idea of carbon trading amongst nations—given their relative industrial capacities and demands and the relative poverty of those less industrialised countries—, Attfield asserts a sage note of caution with regards the real practical and indeed moral advantages of carbon trading. The latent danger of carbon trading is the risk it poses to the very necessary future development of poverty-stricken countries. The reduction of Third World debt is of no benefit whatsoever to the poverty-stricken individual if it comes at the cost of real, positive development on the ground. The worry is that developing countries will trade their emissions rights at the expense of necessary developmental emissions. Attfield's line is that the kind of emissions involved in raising standards of living for the world's poorest must be secured against carbon offsetting.

With these considerations in sight, Dower presents three alternative means of apportioning environmental atmospheric entitlement in which both he and Attfield find some more or less propitious approaches. These include Meyer's "Contract and Convergence" approach, wherein the primary goal is a gradual contraction of global emissions and a progressive convergence of national emissions entitlements, and the proposals of Greenhouse Developments Rights, where greenhouse gas mitigation and development for the satisfaction of basic needs in poorer countries are deemed the shared responsibility of all people at or above a certain level of financial security. The great benefit of this approach, emphasised by Attfield and later developed by Dower, is the shift in moral responsibility from the conglomerate nation to the individual agent. Herein one finds the kind of loop-hole whereby wealthy residents of poorer countries are shielded from the limitations imposed upon citizens of wealthy, carbon profligate countries helpfully curtailed (Dower, 19).

Drawing upon Attfield's arguments, Dower provides compelling justification for Attfield's approach to environmental ethics, but in the second part of his discussion Dower moves towards a more reflective consideration of the ethical and the pragmatic considerations at work within these different principles. From here Dower presents a minor, albeit more than academic distinction between Attfield's thinking and his own. Earlier on in the paper, Dower recognises Attfield's astute exemplification of the way in which different ethical commitments determine the tenor of one's practical solutions (Dower, 17). But whilst Dower recognises the ethical coherence behind Attfield's mode of theorising, he is wary of over-discriminating between attitudes which are practically, but not necessarily theoretically sympathetic to one's own ethical position. Proving himself a keen pragmatist, Dower's point is that we must prioritise the common practical goals of disparate ethical positions in spite of possible disagreements concerning the theoretical rationales behind them. Whilst we cannot disregard these ethical rationales altogether, we cannot afford to alienate sectors of the environmental conversation or to foreclose dialogue and co-operation amongst the disparate but more or less practically united community of environmental ethicists (Dower, 21-24). In short, different ethical positions may foster convergent practical benefits, and it is practical consensus alone which promises the brightest future for our environment. For this reason, Dower affirms the need to draw-out the pragmatic convictions reflecting our ethical distinctions at the theoretical level, all the while developing this argument in the context of two "quintessential 'Attfield'" commitments, namely to non-human species and to future generations (Dower, 18).

In the final instance, Dower draws our attentions to Attfield's treatment of "mediated responsibilities", to the fact that morally significant, in this case environmentally compromising actions, are very often carried out on our behalf at a remove, by the power companies who provide our electricity say, or the governments who establish enabling contracts and statutes with these companies or indeed other nations (Dower, 26). Touching upon the issue of democratic representation and the paradox of the undemocratic electorate—is the undemocratic policy in fact democratic if democratically sanctioned?—Dower ends by addressing the ethical responsibilities of the individual, specifically in relation to national governmental policies. Finally he makes what he calls a "constructive amendment" to Attfield's practical, and in Dower's opinion overly state-centric stance, arguing for the ethical benefits of a more individualistic, less nationalistic or centralised approach to the environment (Dower, 31). For Dower it is the vanguard minority of individually committed agents, groups often deemed radical in contrast to the prevailing majority, who possess the greatest power to transform public opinion and effect the swiftest and most profound change to environmental attitudes and habits of consumption.

In Chapter Four, Alan Holland analyses some of Attfield's key claims in *Creation, Evolution and Meaning* (Attfield, 2006). In Part I, Holland discusses what he calls "[Attfield's] re-working of the cosmological argument" (Holland, 41); a re-working which involves an appeal to the principle of sufficient reason (as outlined by Nozick, 1981, 140-142), which, for Attfield, states that "there is a sufficient reason for whatever could have been otherwise, except where there is a sufficient reason to the contrary" (Attfield, 2006, 57, quoted by Holland, Part I, 41). Holland interprets Attfield as follows: if one takes into account the principle of sufficient reason, an explanation is required for the existence of the material world; for this existence is something that could have been otherwise (there could have been nothing rather than something), and God as the creator stands as an explanation for the existence of the material world. Holland's focus in this section is on the supposition that an explanation really is needed for the existence of the material world. In line with Nozick's principle, Holland suggests that there may be sufficient reason for supposing the contrary, and after clearly setting out the premises of his argument he concludes that "there is no reason for supposing, and every reason for not supposing, that the existence of material objects, per se, requires explanation in terms of something other than themselves" (Holland, Part I, 42).

Holland goes some way with Attfield in accepting that belief in creation and evolution are compatible. However, Holland makes a distinction between two senses of Darwinism: Darwinism as the belief in evolution, and "methodological Darwinism" as a method of science which places significant importance on the providing of evidence. Holland questions whether methodological Darwinism is compatible with belief in a creator. It would seem that methodological Darwinism would require "the admitting of evidence" (F. Darwin 1902, 57, quoted by Holland, Part II, 44), and it is not clear that there is sufficient evidence of a creator.

In Part III, Holland discusses Attfield's re-working of Keith Ward's "argument from value" (Ward, 1982, 89-120), and in particular discusses the claim that "design on the part of a purposive and loving creator is a reasonable interpretation of the immense amount of value in the world" (Attfield, 2006, 151, quoted by Holland, Part III, 45). Holland analyses what he sees as two assumptions underlying this claim: firstly, that value judgments are objective and, secondly, that the amount of value in the world is something which can be estimated. In relation to the first supposed assumption, Holland offers a preferred perspectivist account of value (Holland, 45-46). In respect of the second supposed assumption, Holland draws our attention to (among other things) the difficulties involved in weighing up the value in the world and involved in the claim that there is an immense amount of value in world (Holland, 47-48).

Holland praises Attfield's response to a related objection against the belief in a good God; the objection being that the evil in the world casts doubt on the plausibility of such a belief. Attfield claims that "we have no reason to believe that a world with a better balance of good over evil than the actual world is possible, or that the actual world is not a world that a good God would create" (Attfield, 2006, 141, quoted by Holland, Part III, 47). While Attfield does not necessarily have to show that the world as we know it is the best possible world (something Holland recognises), Attfield does indeed try to show this by pointing out that some evils could be seen to be necessary parts of a natural world that generates value. Although Holland believes that Attfield's case here is convincing, he argues that "the belief in a beneficent creator does require us also to believe both that there is a preponderance of value in the world, and that no greater preponderance could possibly have been brought about" (Holland, Part III,

47), and that we have no adequate reason to believe that this is indeed the case.

Holland criticises Attfield's claim that evolution is purposive in its creation of value in the world (a claim which may further suggest that the value in the world can be seen to be the purposive result of a creator). In contrast to Attfield, Holland argues that it is possible that the world and all its value came about in a non-purposive way; such as by chance and random events and, if this is the case, the value in the world cannot be explained by reference to design (Holland, Part IV, 49).

Finally, Holland discusses Attfield's view that meaningful life could be best achieved through belief in a creator; a creator that has generated the valuable states of affairs that exist in the world. In particular, recognition of ourselves as stewards of all that has value in the natural world can enhance the prospects for a fully meaningful life. Holland objects to Attfield's view here and argues that, in the light of objections outlined in Part III, the prospects for a fully meaningful life are "hostage to a value calculus of cosmic proportions"; a calculus which is "beyond… human capacities" (Holland, Part V, 50). Holland further argues that belief in a creator, far from enabling us to achieve a fully meaningful life, "casts the possibility of such a life into the greatest of doubt" (Holland, Part V, 50).

Creation, ethics and environment are the dominant themes of this book and it is in the sixth chapter by Christopher Southgate that creation, invested with its full theological resonance, comes to the fore. It is important to note straight away that neither Attfield nor Southgate question the fundamental coherence of divine creation and evolutionary development; such is the nature of evolutionary theodicy named in Southgate's title. For the sceptic of course, there is no greater challenge to the presumed benevolence and clemency of God than the endemic suffering of the Darwinian order; disease, predation, famine, and in the most extreme cases total species extinction, are the necessary conditions of evolutionary flourishing on a global scale. The beauty, ingenuity and diversity of creation feeds upon the carcass of an incalculable and irrepressible suffering, does it not? What God, whose God, could warrant such misery? To discourse upon evolutionary theodicy is thus to discourse upon natural ethics, upon the existence of natural evil and the potential balm of a greater good. But natural suffering, of which Southgate treats in the earlier sections of his paper, is no longer the full story. To this difficulty we must add the more timely evils of man-made environmental damage. No longer is it possible to pose the question of divine creation without due consideration of man's role within the course of natural history. If creation involves design, and is a theme of theodicy, where does humankind's destructive impulse, its propensity for environmental evils, fit within the divinely sanctioned natural order? Southgate's proposals concerning the problems of natural, evolutionary suffering and human defilement are sympathetic to Attfield's approach but they are not however identical.

To emphasise the common ground between Southgate and Attfield we may begin by emphasising their shared criticism of the kind of "dualistic" interpretations of evolution and divine creation, whereby God would represent a power extrinsic to evolution itself, and a power moreover that is only present at certain critical and apparently inexplicable junctures within the evolutionary process. Both thinkers reject the notion of God as a kind of supernumerary catalyst or information encoder, opting instead for a vision of God's consistent reciprocity with nature. But this reciprocity, in Southgate's own words, "greatly intensifies the problem of theodicy" insofar as it "makes God not only the author of a process to which vast quantities of suffering and extinction are intrinsic, but it also posits that God is efficiently active in the process without commuting that disvalue" (Southgate, 63). Attfield's and Southgate's respective handling of the matter forms the common ground from whence their opinions begin to diverge.

There is a great deal of nuanced argumentation within Southgate's paper and he draws upon a compendious knowledge of other people's arguments within the field. For the sake of brevity however, it is as well to emphasise the two critical pivots upon which the paper turns. The first of these, alluded to in the "Pangloss" of Southgate's title, concerns Attfield's "only way" argument; this is the thesis that whilst no evidence exists to the contrary, we must assume the actual laws of nature and the less than ideal conditions of evolutionary development to present the only possible means for creation. In fact, Attfield's argument extends to the existence of moral evil, to the evil of human actions within the world as well as the natural evil of evolutionary suffering. We are told that according to Attfield, the "only possible way" view does not contravene the argument for moral agency amongst individuals. That the world could not be otherwise does not mean that there is not free-will amongst God's creatures. It follows for Attfield that moral evil is a consequence of free will, whilst natural evils present "systemic preconditions of the flourishing of billions of creatures across the ages, as well as of human capacities and of the human endowment" (Attfield, 2006, 143, quoted by Southgate, 64).

Southgate's discussion of Attfield's "only possible way" thesis, and of his own reservations regarding it, enables him to delineate their respective modes of enquiry with economic clarity. This is because the questions occasioned by the "only possible way" thesis cut to the very heart of Attfield's ethical commitments as a philosopher. What is more, they help to disclose something of the continuity between Attfield's philosophical biocentrism and his theological position *vis á vis* evil and suffering.

It is Southgate's ultimate conviction that the validity of Attfield's position would be best served within the wider, and he claims "richer" context of eschatology; within the purview of an as of yet unrealised divine redemption or salvation, wherein Attfield's prescription of the "only possible way", and all wider defences of God's justice, can be placed within the context of future good. Attfield, Southgate claims, provides a "thin defence" of God's justice, focussed the latter feels, rather too much upon God's disposition as a kind of "consequentialist calculator of values against disvalues" (Southgate, 66). Attfield may not agree with this characterisation certainly, but it is true to say that Southgate's emphasis upon redemption, and what is more, upon the co-suffering and co-redemption of God and humans—God's co-suffering with humans and humans' co-redeemerly responsibilities to God—endows the question of divine justice, of God, suffering and necessary evil, with a singularly immediate, and if one is allowed to use such terms in this context, humane or humanistic countenance. By drawing attention to the divergence of their theodicies, Southgate proceeds to elaborate the different environmental ethics arising from this difference of opinion concerning the status of moral and natural evil. Where Attfield's biocentrism and "thin defence" of God's justice leads him to adopt an ethic of stewardship or guardianship with regards the natural world, a position focussed upon humankind's duty to protect the natural environment as God's "trustees", Southgate states his preference for what he terms a "co-redeemerly" ethic. Here the emphasis rests upon the shared work of God and humankind, and the latter's role within that work. Here God's call to His people is evinced squarely in terms of humankind's responsibility to the planet, not for the revival of a lost Edenic ideal, but in the prospect of future hope. But this distinction by no means marks an irreconcilable difference between Attfield and Southgate on the ethical and environmental plane. Southgate accordingly ends by confirming the validity of stewardship discourse within the context of climate change.

Concentrating on Attfield's *Value, Obligation and Meta-Ethics* (1995), in Chapter Eight Roger Crisp examines Attfield's view about the scope of moral standing extending to non-sentient living individuals such as plants, and analyses Attfield's idea of the good of beings, understood in terms of their flourishing or wellbeing. Crisp's focus is on what he sees as Attfield's perfectionist idea of the good. Attfield's view, he claims, is perfectionist in so far as "the good of any being consists in its "perfecting" those essential qualities that make it the kind of being it is" (Crisp, 88). Crisp claims that a cultural evolutionary explanation may account for some intuitions about the view that moral standing extends to non-sentient beings: "Plants are invaluable for human beings. Without them, we could not exist, since they contribute essential elements to our environment and our health. We gain great enjoyment from eating them and in some cases from contemplating them. So it is easy to see how a group of beings that placed special value on plants might do better than a group which did not and was prepared to sacrifice them for any short-term benefit. An account of cultural evolution, in other words, may have the resources to debunk the intuitions underlying Attfield's perfectionism" (Crisp, 90). Partly by virtue of an adapted hypothetical scenario, Crisp proceeds to demonstrate why he believes talk of the good of non-sentient beings to be insupportable, and argues for a welfarist conception of the good as a notion which involves talk of what is "good for" some individual being; a notion which rests on "what it would be like to be such a being" (Crisp, 91). Thus Crisp presents an idea of the good which, unlike Attfield's, rules out non-sentient beings as candidates for moral standing.

Crisp proceeds to pose a problem for perfectionist positions (one of which he attributes to Attfield). As he claims "According to the perfectionist, we should first seek an impartial, "value-free" account of a being's essential and species-specific nature. But there is a serious danger of the perfectionist allowing her conception of well-being to guide her account of nature. Consider, for example, the remarkable number of perfectionist philosophers over the centuries (Aristotle among them) who have claimed that it is part of our nature to philosophize!" (Crisp, 91-92). He is then wary of the claim that one's good or well-being depends upon fulfilling what is thought to be in one's nature (Crisp, 92). Crisp suggests that a hedonist theory of value, which could be seen as one in which the good is seen in terms of that which increases pleasure, or diminishes pain and suffering, should be considered as a preferable alternative to a non-hedonistic one when approaching questions about the extent of moral considerability and the idea of the good.

Finally, in Chapter Ten Clare Palmer discusses Attfield's views in respect of sentient animals and their moral standing, and what she calls Attfield's consequential "capacity-oriented" approach to animal ethics: "Attfield's view… is essentially *capacity-oriented.* Moral standing, and what constitutes ethical practice, is based on the flourishing that can flow from the intrinsic capacities that living beings have, independently of these beings' relations to others" (Palmer, 108). Palmer argues that a capacity-oriented approach to animal ethics poses problems for the consequentialist and that this approach does not take sufficient account of relational concerns in determining our moral obligations to sentient animals. There are some relationships that are morally significant, particularly where humans have created relations between themselves and animals in which the animals concerned are dependent on humans for their flourishing. She argues that in considering such relationships and the responsibilities they create, capacity-based consequentialism falls short, partly because of the forward-looking nature of consequentialism and partly because properly considering relational responsibilities requires recognising factors other than consequential ones. But in Chapter Eleven, Attfield suggests that consequentialism, forward-looking as it is, can underpin backward-looking practices such as caring for animals for which one has assumed responsibility.

The recognition of the interests of present, future and prospective individual creatures and the good of all sentient and non-sentient living beings; such are the hallmark traits of Attfield's consequentialist biocentrism. It is nonetheless reassuring to find our contributors presenting Attfield with the range of critical counterpoints that they do. For it confirms that whilst mainstream society may falter over past assurances to safeguard the environment—in spite of an increasingly gloomy and complex picture of environmental damage—the discourse of environmental ethics itself shows no signs of abating or retrenching from the positive commitments, enthusiasms and convictions of its early adherents.

References

Robin, Attfield, 1995, *Value, Obligation and Meta-Ethics* (Amsterdam and Atlanta, GA: Rodopi).

—. 2006, *Creation, Evolution and Meaning* (Aldershot: Ashgate Publishing Ltd).

Darwin, Francis, 1902, *Charles Darwin: His Life* (London: John Murray).

Nozick, Robert, 1981, *Philosophical Explanations* (Oxford: Clarendon Press).

Ward, Keith, 1982, *Rational Theology and the Creativity of God* (Oxford: Blackwell).

CHAPTER ONE

CREATION, ENVIRONMENT AND ETHICS: SOME CARDIFF-BASED CONTRIBUTIONS TO PHILOSOPHY

ROBIN ATTFIELD

I have been asked to say a little about my research contributions to Philosophy, and am attempting here to comply with that request. (References in this chapter to writings of my own correspond to the list of publications presented elsewhere in this volume. A short bibliography of works of others referred to in this chapter can be found at its end.)

Philosophy students often think of their lecturers as teachers only, rather than as researchers too. My very first students would not have been too far out to think of me in this way, as I had until then published nothing in philosophy, although I had been doing philosophy research in the previous few months at the University of Manchester. That research was continued as my doctoral dissertation, which was entirely composed when I was already in post at Cardiff, having been appointed in January 1968. It was inspired by the work of Dietrich Bonhoeffer on secularisation, and concerned secularisation among philosophers from Bacon to Kant, with some amount of philosophy of religion and history and philosophy of science thrown in. The main theme was that theists can welcome secularisation (for example, critiques of religious claims, together with the practice that generates them) as opposed to secularism (a case of which would be the claim that science explains everything and thus leaves no room for religious belief). A second theme was that the critiques supplied by Hume and Kant of defences of theism from contemporary writers such as Samuel Clarke were inconclusive.

After the doctorate had been awarded in 1972, I tried for several years to find a publisher for a book that would be based on it, but succeeded only when University College Cardiff Press came into being in 1976. With them I published *God and The Secular* (1978), a slightly scaled-down version of the doctorate; UCCP insisted that the text be shortened by one sixth. Some thirty years later I was able to publish one of the omitted sections, on secularising critiques made of each other by Leibniz on the one hand and Clarke and Newton on the other (2005, actually published 2007). *God and The Secular* received several reviews, including a strongly adverse one from M.A. Stewart, who was understandably annoyed by shoddy work on the part of the publishers, who had sent out a review copy parts of which were printed upside down. Stewart, however, had a reputation for astringent reviews, and a few years later, when I met him, he retracted his adverse remarks, although only in a private conversation. Many years later, David Lamb arranged the republication of this book in 1993 under an Ashgate imprint called Gregg Revivals.

By this time, I had published some ethics articles in prominent journals, including 'On Being Human' in *Inquiry* (1974), 'Against Non-Comparabilism' in *Philosophy* (1975), and 'Towards a Defence of Teleology' in *Ethics* (1975). ("Teleology" was the term then in use for consequentialism.) I had also published shorter pieces on Berkeley (1970) and on Hume (1971) in *Philosophy*, and on religious language in *Sophia* (1970, 1971, 1972) and *Religious Studies* (1973, 1977), which happened to catch the eye of a later referee, Frederick Ferré, but the ethics articles served as the initial steps towards a later and larger project in that field. The following year, a reply was published to my consequentialist account of justice in *Ethics* (1975) by one Louis Kort, and my colleague Barry Wilkins called round and encouraged me to compose a response, telling me at least what kind of thing I should say. So I took him up on this, and published a short rejoinder, 'Racialism, Justice and Teleology' also in *Ethics* (1977), explaining how justice and consequentialism could be reconciled.

Around this time I had two short pieces in *Analysis*, one on collective responsibility (1971), and the other on the analysis of existence on the part of my former teacher, G.E.L. Owen, who used to write about the presocratics, Plato and Aristotle. This was, for a long time, my only published contribution to Ancient Philosophy, and it was an indirect one at that. It was called 'How Things Exist: A Difficulty' (1973). (Michael Durrant tells me that he later came across a reply to it in someone's doctoral thesis.) Michael Durrant also collaborated with me in a joint article, published in *Noûs*, called 'The Irreducibility of 'Meaning'' (1973), which was my only foray into that field until the publication of *Creation, Evolution and Meaning* (2006). In this paper, we resisted analyses of meaning in terms of use, or more specifically of illocutionary act potential, on grounds of their circularity. Michael was very much the senior partner in this enterprise, but showed great fortitude both in face of repeated requests for revisions from the editor of *Noûs*, Hector-Neri Castañeda, and of intrusions on his time and attention from me, one of them when he was in hospital.

I spent the academic session of 1972-3 teaching at the University of Ife in Nigeria, taking with me not only a wife and a young family but, more relevantly to research in philosophy, a document called 'Forty Related Views'. These forty theses were positions in and around ethics, such as naturalism and consequentialism, which I would have liked to write up in book form. But at that stage I had so much difficulty finding a publisher for *God and The Secular* that this project had to be deferred for several years. Instead I wrote up a by-product of my dissertation, the paper 'Clarke, Collins and Compounds', published in *The Journal of the History of Philosophy* (1977), which later made me one of their readers. There was also an article in *The Journal of Religion* on 'Science and Creation', concerning the paradoxical relations between science and that doctrine (1978). In the late seventies I had a paper on supererogation accepted by *Mind* (1979), and submitted with similar success a paper on moral relativism to *The Monist* for one of their special numbers (1979). There was also a paper on inter-generational reparations, published in the Nigerian journal *Second Order* (1979), and earlier presented to the Cardiff Philosophy Seminar.

While in Nigeria I composed a calypso about De Morgan's Law, suitable to be sung to the tune of 'Immortal, invisible, God only wise'. It runs:

Not either, so neither,
De Morgan he say,
And also vice versa;
It works either way.
Not both, so one only
At most, this he teach;
And also conversely
Just one, so not each.

The Nigerian students, at least, seemed to like this and were seen and heard singing it lustily.

By the late 1970s I had become interested in environmental philosophy, and soon wrote my first paper in that field, 'The Good of Trees', first published in *Journal of Value Inquiry* (1981), and later much anthologised. Sources of this new interest included the influence of Henry Odera Oruka, a colleague at the University of Nairobi, where I spent four months teaching in 1975. Oruka had attended the World Congress of Philosophy at Varna, Bulgaria, at which he came across one of the earliest papers ever written in this new field, Richard Routley's 'Is There a Need for a New, an Environmental Ethic?' (Routley, 1973); this whetted my appetite. Another source was my reading (on my return to Cardiff) of John Passmore's book *Man's Responsibility for Nature* (1974), to an earlier draft of which Routley had been replying. While endorsing Routley's rejection of anthropocentrism, I was attracted by Passmore's approach to Western ethical traditions vis-à-vis nature. Yet more compelling a source was being asked to teach in 1977 one term's worth of lectures towards the Cardiff course 'Philosophical Aspects of Social and World Problems' in place of Vernon Pratt, who had moved to Lancaster University in 1976. (This course had been approved on a split vote in the Faculty of Arts in 1973, while I was still in Nigeria, and marked the first stirrings of a Cardiff involvement in applied ethics.) Having to teach a field fosters both puzzlement and the adoption of stances within it. And so, before long, deferring plans for writing a work on ethics, I began composing what became *The Ethics of Environmental Concern* (1983), which, as I correctly guessed, was more likely to find a publisher.

This book combined distinctive contributions to the field of environmental ethics, including a defence of a consequentialist version of biocentrism, with a discussion of population ethics owing much to Derek Parfit's (as yet unpublished) work (later published in his *Reasons and Persons* (Parfit, 1984)), together with a critical survey of Passmore's account of Western religious and ethical traditions. In this connection I followed up Passmore's favoured source, Clarence Glacken's *Traces on the Rhodian Shore* (Glacken, 1967), and rapidly found that it supported an account far removed from that based on it by Passmore. At the same time, I was able to redeploy Passmore's classification of what he regarded as "minority traditions" such as that of stewardship, and to cite many of Glacken's sources to show that it was far from the recessive tradition of Passmore's account. At one point I planned an even ampler book, which was to have included a section on environmental politics, but my colleague Professor Thomas McPherson advised me to treat the book as complete as it then stood; and no one has ever objected to this omission. Reviewers ranged from those claiming that this was a survey book (admittedly part of the truth, for it was effectively the second overview of the field, Passmore's having been the first) to those hailing it as deserving an "environmental Oscar". It soon led to my being invited to chair a section of the Montreal World Congress of Philosophy of 1983, and in 1984 to my being invited to join the Environmental Ethics Working Party of the Ian Ramsey Centre at Oxford University.

There were three substantial papers published separately as by-products of this book. 'Christian Attitudes to Nature' (1983) was a reply to White, who had ascribed anthropocentric and despotic attitudes towards nature to Christianity in a paper in *Science* called 'The Historical Roots of Our Ecological Crisis', and also to Passmore, who had rejected some of White's views and accepted others. It was published in *Journal of the History of Ideas*, and was apparently made prescribed reading in some universities. Jan Decker told me recently that reading it had saved him needing to look up the sources at the time. (Much more recently I published an essay about White in *Environmental Ethics* (2009), which delved into 'The Roots of Roots' (as one section is called).) A second essay was 'Western Traditions and Environmental Ethics' (1983), and discussed historical attitudes to animals. It was published in an Australian collection, but was badly mangled by one of the editors, as emerged from the University of Queensland version. However, I managed to persuade the Open University Press, which published the British version, to include an acknowledgement of the *errata*. The third was 'Methods of Ecological Ethics' (1983) in which I defended reasoning outwards from agreed cases, an approach that most environmental philosophers condemned as "extensionism". This was published in *Metaphilosophy*. (All three essays later appeared together in my Ashgate collection *Environmental Philosophy: Principles and Prospects* (1994).) Another related talk, 'Christianity and the Ecological Crisis', related this research and the themes of *God and The Secular*, and was broadcast on Radio 3 in 1981.

Other papers of the 1980s concerned very different themes. I wrote on the analogical nature of theological predicates in 'Religious Symbols and the Voyage of Analogy' (1980), which somewhat over-used the theme of Scylla and Charybdis, but was still published in the *International Journal for Philosophy of Religion.* My interest in this topic was kindled by my former colleague Humphrey Palmer's book *Analogy* (1973). Some years later I published a second (and more readable) essay in this field in the *Scottish Journal of Religious Studies* (1996). Several papers about Leibniz appeared in *Studia Leibnitiana* (1980), reflecting themes from my dissertation, which were differently echoed in articles and reviews about religious language that appeared in *New Blackfriars* (1980, 1983, 1984); I don't know whether any other Quakers have ever written for that journal. There were also some papers on ethics, such as 'Optimific, Right but not Obligatory' (1982), published in the *Canadian Journal of Philosophy*, 'Work and the Human Essence' (1984), which appeared in the first number of *Journal for Applied Ethics*, and another joint paper with Michael Durrant, critical of the communitarianism of D.Z. Phillips and H.O. Mounce, which appeared in *Philosophical Papers* (1981). (Most of the credit for this paper belongs to Michael, who supplied most of the ideas; my role was to make their presentation more reader-friendly.) Another applied philosophy paper was 'Development: Some Areas of Consensus' (1986), which the editor of *Journal of Social Philosophy* declared, in accepting it, "almost the ideal paper" for that journal. In this period I was also able to resuscitate and revise some research on the impacts of secularisation on the history of witchcraft, that I had done at Manchester before coming to Cardiff, which was now published in the journal *Annals of Science* (1985).

A light year of teaching now gave me the opportunity to write up 'Forty Related Views' as a book which Croom Helm agreed to publish as *A Theory of Value and Obligation* (1987), covering value-theory, normative ethics and meta-ethics. It was probably a mistake to cover all these areas together, but that had long been the whole idea. There were quite a number of reviews, some of them quite favourable, but tending to say that, since nothing I or anyone else could say would persuade non-consequentialists to become consequentialists, its main interest consisted in the modifications to ethical theory that environmental ethics seemed to require. There was also a scathing review from Frederick Feldman, and an extended and largely favourable one from David O. Brink in *The Philosophical Review* (1991). These and the other reviews were among the factors that impelled me to rewrite the book, which re-appeared from Rodopi of Amsterdam and Atlanta as *Value, Obligation and Meta-Ethics* (1995). That book received fewer reviews, but responses eventually included some largely favourable coverage from Alan Carter, effectively itself a review, presented in a review of a later book of mine in *Mind* (2001).

Meanwhile I had been rather distracted from philosophical research for parts of the 1980s because of the struggle to reform the structures of University College Cardiff, and limit the powers of the then Principal, C.W.L. Bevan. From 1984 I was also spending one day a month with the Environmental Ethics Working Party of the Ian Ramsey Centre at Oxford. Eventually I became the joint editor of its report, *Values, Conflict and the Environment* (1989 and 1996). That report advocated a method for environmental decision-making called "Comprehensive Weighing", an environmentally enhanced version of cost-benefit analysis, which I agreed to support as a compromise, although I spent some of the next few years presenting papers explaining my reservations and objections (1995). As for the report, it was published (1989) by the Cardiff Centre for Applied Ethics (together with the Ian Ramsey Centre, which was at that time out of funds); the report was republished a few years afterwards (1996) by a more established publisher, Avebury (the present Ashgate). Another result of my participation in this enterprise was that in the dark days of the late 1980s, when the future of the Cardiff Philosophy unit was in the balance, Lord Bullock, the co-chair of the Oxford Working Party and former Vice-Chancellor of Oxford University, wrote to the Cardiff Principal, Aubrey Trotman-Dickenson in our support.

In the early nineties, some of the participants in the Working Party arranged a conference at the University of Kent at Canterbury in which we applied this method to an actual planning decision (about road-building in Dorset), coming up with the conclusion that a different decision from the one reached by the Inspector would have been better, as also would have been doing nothing at all. The report continued to be discussed occasionally, but otherwise fell into oblivion. Meanwhile the view shared by Alan Holland and John O'Neill that cost-benefit analysis was incapable of being reformed prevailed, at least among philosophers.

In the early nineties, Barry Wilkins and I jointly edited for Routledge *International Development and the Third World: Studies in the Philosophy of Development* (1992), each contributing an essay. Mine was 'Development and Environmentalism' (1992 and 1994), which argued that developmentalists (that is, supporters of social and economic development) should support environmentalism, the cause of environmentalists, and also vice versa. This essay won high praise from Susan J. Armstrong and Richard G. Botzler in one of their survey books (Botzler and Armstrong, 1998, 93). Then, in 1993, Andrew Belsey and I organised in Cardiff the Annual Conference of the Royal Institute of Philosophy, on the theme 'Philosophy and the Natural Environment', and the next year edited the conference proceedings under the same title. My own paper was entitled 'Rehabilitating Nature and Making Nature Habitable' (1994). Andrew and I wrote the Introduction. However, I was distracted from composing articles through serving a stint as Chair of the Philosophy Board of Studies, from 1991 to 1994, until I was granted a year's study leave in compensation. It was in that year that *Value, Obligation and Meta-Ethics* (1995) was prepared.

Across the nineties, a steady flow of invitations came my way to present papers, some in Kenya, some in Malaysia, one in Austria, one in the then Czechoslovakia, and several in Scandinavia. (The invitations to Kenya were from Henry Odera Oruka, now a Professor, and a prolific organiser of conferences until his untimely death in the later part of 1995;

his 1991 Conference was a World Conference of Philosophy, sponsored by FISP, though lacking the status of a World Congress.) There was also a trip in 1994 to present a paper at the American Philosophical Association (Pacific Division) at Los Angeles, which somehow generated invitations to present other papers at Corvallis in Oregon and at Fort Collins and Boulder in Colorado. Most of the invitations were in the field of environmental philosophy, but I continued to give papers in other fields too, such as the history of philosophy (1993, 2004) (a field in which I served on the Committee of the relevant British Society) and the ethics of genetic engineering (1995, 1998). But there were still opportunities to give talks about most of the range of topics in environmental philosophy, from historical and religious themes, via aesthetic ones, to issues of population policy, sustainable development and the compatibility of environmentalism with democracy. There is no space to cover them all here.

In 1998, after taking part in Nigel Dower's Aberdeen-based global citizenship project, I was awarded a semester of study leave to write a book for his series on Global Ethics for Edinburgh University Press, which emerged as *The Ethics of the Global Environment* (1999). Most of the issues just mentioned were discussed, together with some thought experiments about human extinction, and an updated account of the stewardship approach; this latter I was also asked to present at a colloquium of the John Ray Initiative, held at Windsor Castle (of which I contributed to the proceedings volume (2006)). The book on the global environment was reviewed for *Mind* by Alan Carter (2001), who (as mentioned above) obligingly reviewed *Value, Obligation and Meta-Ethics* at the same time, as conveying its theoretical basis. Carter regarded this theoretical approach as being as good as any, but regarded all such approaches as defective on pluralist grounds. I published a reply in *Utilitas* (2003), to which Carter responded there at length; I was then allowed the last word (2005), albeit a fairly short one. (There is a further unpublished contribution of mine to this debate; but the Paris conference where it was presented has not yet managed to publish the proceedings.)

Discovering that John White had published in a 1997 book a reply to my earlier paper 'Work and the Human Essence', I now published a rejoinder in the first number of *Reason in Practice* (which now carries the title *Philosophy of Management*) (2001). This allowed me to resuscitate a long-standing argument from essential human capacities, which he had misconstrued. This paper was apparently regarded by one of the editorial board as the corner-stone of their first number. But since then, this argument seems to have returned to oblivion.

In 2000, together with Johan Hattingh of Stellenbosch University, I managed to secure funding from the Association of Commonwealth Universities for a project on sustainable development, with reference to South Africa. There followed two visits in 2001 and 2002, one to the Northern Province, and the other to Stellenbosch, to which I was returning, having earlier held a Research Fellowship of the National Research Council, which had sponsored lectures at Stellenbosch, Cape Town and Pretoria. This time my wife Leela became an officially funded member of the research team. Johan also visited Cardiff more than once. The upshot was two papers on sustainable development, one published in *International Journal of Human Rights* (2002) and the other in *Third World Quarterly* (2004). The second of these papers was jointly authored by Johan, myself and Manamela Matshabaphala, an academic at Witwatersrand who had participated in the project. A related (overlapping and therefore unpublished) paper was presented at Addis Ababa in 2005, where Leela and I were visiting the University at the invitation of my former doctoral student, Workineh Kelbessa.

During 2002, with the help of another semester of study leave, I composed a textbook for Polity Press of Cambridge called *Environmental Ethics* (2003). At the suggestion of Patricia Clark I included a glossary, portions of which have often been quoted back at me by student essayists since the book was published. This book includes a further defence of biocentric consequentialism, and applications of it to, for example, global warming. Soon afterwards, I began presenting papers on that topic, one of them at San Diego, which was recently published in a special number of *Journal of Social Philosophy*. It is entitled 'Mediated Responsibilities, Global Warming and the Scope of Ethics' (2009), and stresses the enlarged scope of human impacts that, because of technology, ethics needs to consider nowadays, as not in the past.

Several other papers on global warming have followed, partly because of my participation in a UNESCO Working Party on environmental ethics (2006), and also in preparing the *White Paper on the Ethical Dimensions of Climate Change*, authored largely by Donald Brown of Pennsylvania State University (2006). This seems to have generated an invitation to address an international conference of Fondazione Lanza in October 2008 at Padua, another to address the Toda Foundation conference at Honolulu the next month (to which I contributed electronically), and another to address a conference in March 2009 at the National Leonardo Da Vinci Museum at Milan. A further such paper was presented at the World Congress of Philosophy at Seoul (Korea) in August 2008. (The relevant papers are all forthcoming.)

In 2004, I read an editorial by Alan Holland in *Environmental Values* which came across as a challenge to myself, and composed a reply that was published (2005) in the same journal. Since the challenge concerned the poetry of John Clare, the reply used passages of poetry in an attempt to illustrate that not all talk of value is to be regarded as a projection of human sentiments. I am grateful to Christopher Norris for some of the ideas, particularly a relevant passage from Gray's *Elegy*.

In 2005, I was awarded research leave by Cardiff University and by the Arts and Humanities Research Board (one semester from each) for a book on creation and Darwinism. One of the referees was Frederick Ferré, one of few philosophers able to remember my much earlier writings on philosophy of religion. After planning the structure during a holiday in 2004 to visit a nephew and his family in Vietnam, I managed to compose *Creation, Evolution and Meaning* (2006) by the end of 2005. The front cover design is a photograph that I once took of the Eden Project in Cornwall. The book was published in December 2006, and has so far received several favourable reviews, plus a less favourable one in *The Times Higher*, to which I managed to publish a brief reply by way of a letter to the editor. In this work, philosophy of religion is combined with philosophy of biology and environmental philosophy. The opening chapters result from reflection on the work of Alexander Miller and Christopher Norris on realism and anti-realism, as well as some longstanding interests in verificationism and in falsification. Clarification of belief in creation (and of how to distinguish it from creationism) and arguments for the existence of God are present, as is a discussion of the varieties of Darwinism, its relation to varieties of teleological argument, and its relation to the problem of evil. Later parts discuss Daniel Dennett's account of how the story of evolution culminates in language and then, supposedly, in science, adding how a stewardship approach coheres with the account of creation given earlier. One indirect effect has been an invitation from Christopher Southgate to write reviews for *Reviews in Science and Theology*. I hope to develop further some of the themes about evolution and psychology in a project funded by the Oxford University Ian Ramsey Centre for the year 2009-10.

Once this book was out of the way, I was able to turn to a contract with Ashgate to edit a collection of readings on environmental ethics. My main role was to write the introduction, which occupied much of the summer of 2007; the book (much the longest that I have published) appeared under the title *The Ethics of the Environment* (2008), and has not yet been reviewed, as far as I am aware.

Having embarked on teaching ancient philosophy in a module delivered three times to MA students, I presented a related paper last year to the Cardiff Branch of the Royal Institute of Philosophy. This arose also from the undergraduate module 'Philosophy and Poetry', jointly taught in recent years by Christopher Norris and myself, and was called 'Philosophy on Poetry, Philosophy in Poetry' (forthcoming). It concerns whether Plato would have (or could consistently have) banned philosophical poetry, such as the poetry of Lucretius, from his Republic. It is being published in a proceedings volume of the Society for Value Inquiry sessions held at the Seoul (Korea) World Congress, where I also presented it last August.

Currently I have a contract to write a textbook on ethics, with a deadline of June 2011, and am all too likely to pick the brains of colleagues in the coming two years while I am attempting to write it. No other books are planned, but several conference presentations are envisaged. Having now covered my books and a fair sprinkling of my other writings, I will close, with many thanks to colleagues past and present whose ideas and encouragement have made these writings possible, or (in some cases) who have joined in writing them.

References

Attfield, Robin—see the publication lists at the end the book.

Berry, R.J. (ed.), 2006, *Environmental Stewardship: Critical Perspectives—Past and Present* (London: T&T Clark/Continuum).

Botzler, Richard G., and Susan J. Armstrong, 1998, in Richard G. Botzler and Susan J. Armstrong (eds), *Environmental Ethics: Divergence and Convergence*, 2nd edn. (Boston, MA: McGraw-Hill).

Brink, David O., 1991, Review of Robin Attfield, *A Theory of Value and Obligation*, in *The Philosophical Review*, 100, 140-148.

Brown, Donald, et al., 2006, *White Paper on the Ethical Dimensions of Climate Change* (Philadelphia: Rock Ethics Institute).

Carter, Alan, 2001, Review of Robin Attfield, *The Ethics of the Global Environment* (1999), in *Mind*, Vol. 110, 149-53.

Glacken, Clarence, 1967, *Traces on the Rhodian Shore* (Berkeley, CA and London: University of California Press).

Palmer, 1973, *Analogy* (London: Macmillan).

Parfit, Derek, 1984, *Reasons and Persons* (Oxford: Oxford University Press).

Passmore, John, 1974, *Man's Responsibility for Nature* (London: Duckworth).

Routley, Richard, 1973, 'Is There a Need for a New, an Environmental, Ethic?', *Proceedings of the XVth World Congress of Philosophy*, Varna, Bulgaria: XVth World Congress of Philosophy, 205-210.

White, John, 1997, *Education and the End of Work: A New Philosophy of Work and Learning* (London: Cassell).

CHAPTER TWO

ROBIN ATTFIELD: CHANGING THE ETHICAL CLIMATE ON CLIMATE CHANGE

NIGEL DOWER

Abstract

In this paper I examine Robin Attfield's defence of the idea of an equal entitlement to the atmospheric commons, his rejection of two historical approaches (Kyoto 1990 levels basis and aggregate emissions), and his neutrality between the Contract and Convergence approach supplemented with Millennium Development Goals commitment and the Greenhouse Development Rights approach. In the discussion I distinguish between the diversity of moral theories that support his principle, suggesting a widening of these, and the plurality of working principles that may be needed, again suggesting that the range may need to be broadened. Using another article on mediated responsibilities, I argue that the main emphasis on what states ought to do needs to be supplemented with a robust account of what individuals—in both the North and in the South—ought to be doing, prior to and independent of what laws, taxes, markets or social mores require—both as morally required in itself and as a necessary condition for states effectively doing what needs to be done.

Introduction

I have chosen this aspect of Attfield's wide-ranging interest in environmental and related matters, partly because it is something he has focussed on in a number of writings in recent years, and partly because it is a matter on which I want to develop my own thinking.

It will come as no surprise that I am in large measure in agreement with Attfield's approach to this vital issue. It was not apparent at the time when we first met as undergraduates reading Greats at Oxford in the early 60s that our interests would develop in similar directions, but by 1977, when as I recall Attfield invited me to Cardiff to give two papers, we realised that we had developed similar interests in the environment and development issues, we had both developed a global or cosmopolitan perspective, and what's more we had both got involved with Quakers—and both subsequently became members of the Religious Society of Friends.

It was on the basis of our similarity of approach, and my acquaintance with Attfield's writings, not least his *The Ethics of Environmental Concern* (Attfield 1983), that I invited him to contribute to the series I edited on world ethics and he wrote *The Ethics of the Global Environment* (Attfield 1999). In this he was already tackling the issue of climate change, and later I got him to send several articles which I have used in some teaching I have done since in Iceland and in the USA. In this lecture I am going to take up some ideas in his article 'Global Warming, Justice and Future Generations' (Attfield 2003) and also in an article that is about to be published which Attfield was kind enough to send to me, entitled 'Climate Change: the Ethical Dimension' (Attfield 2008). I shall later on turn to some ideas in an article with a different focus: 'Mediated Responsibilities, Global Warming and the Scope of Ethics' (Attfield 2009).

Part I

Acceptance of the principle of an equal entitlement to the atmosphere's absorptive capacity and rejection of historical approaches

In the first mentioned article Attfield basically argued for the fundamental principle that we should recognise that the absorptive capacity of the atmosphere comprises an instance of the Common Heritage of Humankind and on this basis we need to proportion the amounts of emissions from countries according to population size. This I take it means working out what the average level of emissions for people in the world would need to be in order for increases in temperature to stabilise at an acceptable level, this being generally understood to be the lowest that is at all feasible. This is now generally understood to be stabilisation by 2050 at no more than 2 degrees above the mean temperature at pre-industrial times. (We should note that it has already gone up considerably and that some of the further rise is inevitable because of the emissions already released.) Anyway it is now understood by most scientists (though this figure is not mentioned in Attfield's paper and was not then agreed) to be approximately 80% reduction for industrialised countries. The adoption of this principle would mean drastic cuts in emissions in a country like the UK by 80% by 2050—preferably, as George Monbiot says, by having the major reductions earlier rather than later. To aim at less would be to exceed our entitlement at the expense of others (Monbiot 2006).

Attfield accepts that since some countries, mainly because of their poverty, have average emissions well below the stated levels, there can be some trading in carbon emissions, but rightly argues that this must be done in such ways as to ensure that the meeting of basic needs in poorer countries is safeguarded, and that trading does not merely get used to cancel debts. This is a theme he is keen to stress generally, particularly in the later paper (Attfield 2008), that our commitment to emissions reduction should be consistent with taking seriously our obligations to the world's poor generally, not merely in response to poverty either incurred or increased by climate change.

Reference to trading in carbon emissions may remind people of the language of the Kyoto Protocol, but Attfield is rightly highly critical of the Kyoto Protocol. He accepts that it may have been the best that one could have expected at that time given negotiation realities, but nevertheless it is deeply flawed as providing an adequate basis for carbon reduction. This is because it is arbitrary to choose carbon emissions at any given date—in Kyoto's case 1990—, and because such a process is deeply skewed in favour of rich countries who already had a disproportionate level of emissions in 1992. We can add, as has become even more apparent in later reports from scientists, that it is entirely inadequate as a total level needed. Attfield rightly points out that what we want is a principled basis for reductions, not the choice of an arbitrary historical date.

He also rejects another historical approach, namely the aggregate historical emissions principle. This is that since the industrial countries in the last two centuries have been doing almost all the emitting, they should be penalised for this and make substantial reductions whilst allowing the developing countries a fairer share in the absorptive capacities of the planet for the purposes of development. Attfield rejects this "historical" principle both on pragmatic and on principled grounds. The pragmatic one is that there will be interminable disputes about quantities of past emissions. The more general problem with the principle is that it conflates causal responsibility with moral responsibility, since for a long time carbon emissions were regarded as a part of the natural cycle rather than disrupting it (Attfield 2003: 20). It also allows the past to set the patterns of greenhouse gas emissions for the indefinite future, it does not pay sufficient attention to future needs, and it does not hold to account properly what newly industrialising developing countries may be doing—an issue now of some significance when one looks at the carbon emissions of countries like India and China.

Attfield's move here is reminiscent of a move in another area which indicates his general preference for ethical principles not based on historical factors, namely his reply to Robert Elliot over "faking nature" (Elliot 1992; Attfield 1994). Elliot argued that areas of land that are allowed to revert to an unmanaged state, for instance after logging, are not "wildernesses" since they lack the relevant historical pedigree of never having been interfered with by human beings and thus lack the value of wilderness, whereas Attfield argued that wilderness is constituted by the current character of wildness, that is biota in an area interacting with each other in a wholly unmanaged state, and the value of wilderness resides in these current values. (I believe this to be indicative of a wider cleavage, not always noted, between those who see history as relevant and those for whom the character of current states of affairs are relevant: consider the difference between Rawls' structural account of justice and Nozick's historical account of entitlements (Rawls 1971; Nozick 1974): consider also questions about whether human identity is constituted by current relationships or by history and tradition (related to our positions *vis à vis* communitarianism and the importance of family history for a person's identity).) I am sympathetic to Attfield's approach in both cases concerning aggregate emissions and wilderness, and thus to his wish not to base what we ought to do about climate change on historical factors, as the aggregate emissions approach does.

On the other hand, if we grant that the "polluter pays" principle has relevance and also recognise that it is in a sense a history-based principle which makes our responsibility extend at least as far back as agents *ought to have been aware* of their causal impacts and therefore have some moral responsibility, then quite an important aggregate emissions story can be told in terms what has been happening in the last twenty to thirty years. Attfield notes in the second paper that in practice "the international community can reasonably require the big polluters significantly to reduce their emissions" (Attfield 2008: 3). I will return to this issue later.

Attfield is aware that criticisms can be made of basing allowable emission levels on population levels. Would this not reward countries which expanded their populations and encourage population growth, something which environmentalists argue we have good reasons not to do? This can be dealt with by basing the emissions total on current adult populations or by stipulating that no country could be benefited with population increases beyond an agreed percentage after an agreed starting-date. Attfield also considers a criticism that carbon emissions trading would simply lead to reduction of third world debt and leave existing structures unchanged. His reply is that such a policy would need a proviso that basic needs emissions would be made untradeable as Henry Shue has argued (Shue 1994), or that in other ways the policy ensures that basic needs of poor countries are properly met. Attfield's later paper (Attfield 2008) makes more of this "development" objective.

Finally in this first paper (Attfield 2003), Attfield makes the point, familiar enough to anyone who knows his other writings on environmental ethics, that if independent value is accepted for the lives of nonhuman creatures, and if our emissions policies endanger them, for instance in destroying their habitats, then our principle needs to take this into account as well. In a later paper (Attfield 2009) he argues that taking biocentric considerations seriously will lead to significantly different policies.

Contract and Convergence + Millennium Development Goals or Greenhouse Development Rights?

In the later paper (Attfield 2008), Attfield positions his reflections in relation to a recent *White Paper on the Ethics of Climate Change* (Brown 2006). Attfield was himself part of the group involved in preparing this, though he notes his own position is in some respects different from the official line. The White Paper seeks to locate its position firmly within the framework of principles which have been accepted in UN documents and declarations on sustainable development, such as: the responsibility of states not to cause damage to states outside their jurisdiction; the "polluter pays" principle; the acceptance of common but differentiated responsibilities, with the idea that developed countries should take the lead; and the precautionary principle of not waiting until there is scientific certainty before tackling problems.

Before turning to the question of who should bear the burdens, Attfield asks the question: "who are the beneficiaries?" and his answer is, as partly noted before, "all the affected parties (whether bearers of human rights or not), where affected parties include all creatures, human and nonhuman, of the present and of the future, and where future creatures include all those who in one scenario or another could be brought into being" (Attfield 2008: 3). I quote this in full both for two reasons. First, it is quintessential "Attfield", giving in carefully worded summary form many of the features of his approach to environmental ethics—human and nonhumans, any future generations, caution about rights discourse in this area, and a position *vis à vis* the Parfit challenge. Second, it illustrates how Attfield's ethical basis is not necessarily going to be that of many others, including others involved in producing the White Paper, some of whom may be, for instance, enlightened anthropocentrists rather than biocentrists. I shall return to this issue later.

Attfield recognises that the key issue is finding a sustainable system for burden-sharing. He considers again the aggregate emissions approach, and notes again some of problems with this approach. But he adds the argument that if we are concerned with a sustainable regime, we need to look at what current agents are doing according to the "polluter pays" principle; and, importantly, this must include developing countries as well. These countries may not have been historically involved in causing the problems, but they now need to participate in an equitable greenhouse gases reduction system. "Once developing countries can generate enough electricity to satisfy the basic needs of their populations, this responsibility, it can reasonably be argued, involves these countries in stabilising their emissions rather than relying exclusively on reductions made by currently developed countries" (Attfield 2003: 3).

Turning directly to the issue of international equity considered in the White Paper, Attfield notes that the various approaches are of rather unequal quality, but gives support to the most prominent *principled* approach—that of "equal per capita emissions allocation" based on the idea that all human beings should be entitled to "an equal share of the atmospheric commons". This is what the White Paper favours, along with, as a particular interpretation of this principle by Aubrey Meyer, the contraction and convergence account in which there would be contraction of total emissions moving eventually to convergence of human entitlements (Meyer 2005). This approach is widely supported e.g. by Monbiot (2006) and Singer (2002). There are, Attfield notes, two criticisms of this, one which he regards as a flawed criticism, the other a criticism which whilst valid can be answered with a further qualification.

Some argue that this would put too big a burden on rich countries, so it is preferable that all countries should be expected to reduce emissions, under the Comparable Burdens Principle, by requiring reductions as a percentage of GDP. Rich countries would thus do more than poor countries because they have bigger GDPs. The main ethical problem Attfield has with this—quite rightly in my opinion—is that it would require poor countries to forgo the use of available resources, even if these resources could instead have been used to provide for the unsatisfied needs of their populations.

The other criticism is that the redistributive element would quickly diminish, so the problems of underdevelopment (poverty, malnutrition and disease) would not be addressed, which would be both ethically unacceptable and demotivating so far as poor countries are concerned. Attfield thinks this problem can be dealt with, and he deals with it after he considers an alternative approach. This is called the Greenhouse Development Rights approach (as advocated by Baier, Athanasiou and Kartha 2007) which basically states that the combined costs of greenhouse gas mitigation and promoting development to meet basic needs should be shared by everyone at or above a certain level of development, and all human beings are thus recognised as entitled to development in terms of a quality of life at or above this same level. One implication of this is that "rich people in poor countries would have responsibilities toward funding the total effort, as well as those in developed countries" (Attfield 2008: 5). I give this quotation in full because it includes an important consideration not often noted: that there are, as Paul Harris also argues, many very rich and high-polluting individuals in poor countries as well (Harris 2009). The significance of this, and more generally the importance of what well-off individuals do, is something I want to return to in my later discussion.

Attfield is clearly much attracted to this approach, not least because it includes a serious interest in issues of development. Nevertheless he recognises that the Contraction and Convergence theory could be combined with a serious commitment to the Millennium Development Goals (the 2000 MDGs), and if so, the practical consequences could be broadly similar to the Greenhouse Development Rights (GDR) approach and so similarly ethically acceptable. He appears to be open-minded about which of these to commend, and notes a point, which to my mind would be valid even if he had come down in favour of one rather than the other, that since a strong post-Kyoto agreement is so crucial for humanity, no purist approach insisting on the universal and unadulterated adoption of one approach rather than the other is appropriate, and that more than any one such approach is more likely to succeed.

Part II

Discussion

My discussion will fall into two broad areas. First, I want to look at the theories and principles involved, looking both at what the fundamental ethical principles and theories are here and also at the more pragmatic principles which have a reasonable chance of being used in international negotiations. Second, I want to say something about the pivotal role of individuals in all this—something Attfield does not say much about in these papers but rather more in the third paper I mentioned (Attfield 2009)—since I believe that this perspective perhaps provides a corrective to what seems to be commonly an unduly state-centred analysis of the ethical issues involved. I see this as a constructive addendum to Attfield's approach.

The extent of the plurality of principles: two types of plurality

First I would like to push at Attfield's discussion a little over the issue of levels of pluralism involved. In the first paper (Attfield 2003) he acknowledges that a number of different ethical theories—consequentialism, Kantianism, any theory that is concerned with preserving the enabling conditions for human well-being—can endorse "the equal share in the absorptive capacity of the planet" approach. In the second paper's discussion (Attfield 2008) he acknowledges that both the Contraction and Convergence (CC) approach supplemented with MDG commitments and the GDR approach may both need to be accepted. Here it is worth making explicit that we are dealing with pragmatic principles at a level other than that of theory or general moral principle.

What strikes me about the latter plurality is that they may (but need not) come out of broadly similar more theoretically principled positions. The point is this. Both the augmented CC approach and the GDR approach take it that there are two important moral desiderata: cutting down emissions to one's fair share of the global commons and meeting the basic needs of all people. The difference is about how to operationalise these two commitments. If on the other hand we ask why people might accept these two commitments, it seems to me that at another level we may have a number of different ethical theories that accept them, as Attfield notes. Having made this distinction of levels, I want first to consider the range of theories that might support the ethical principles and the range of pragmatic principles that might implement them; and second I want to raise an issue of interpretation of how to understand the two commitments to curbing carbon emissions and to poverty reduction.

A wide variety of ethical theories supporting the ethical principles

What Attfield says about a number of theories supporting a given ethical stance seems to me to be important not merely in this case, but for the realistic possibility of the emergence of any kind of global ethic as an ethic that is acceptable to a wide range of persons from all over the world. A global ethic as something global in the scope of its acceptance (rather than global in the scope of its content for any given believer in it) is not something universally agreed—that is an idea that is very unlikely to be realised—but rather one that is widely accepted and genuinely available to people from all over the world, and is not subject to cultural bias or ethical imperialism (see e.g. Dower 2005; 2007). The speed of the emergence of such an ethic—as more than a very feeble lowest common denominator—is crucial I think to the project of getting serious cooperation globally on climate change mitigation. So for both practical and theoretical reasons I feel Attfield ought to be more generous about the kinds of theories that could—and will—endorse the fundamental principles of a fair share of the absorptive capacity and commitments to poverty reduction.

I have in mind his scepticism about human rights and contract approaches. I appreciate his concerns about the limitations which these approaches suffer in accounting for our obligations to future generations, but feel that in both cases, such a theory can be so presented as to accommodate this.

I have never been entirely clear why rights discourse, in order to have any grip, has to identify actual living people whose rights are the source of our obligations to them; if future people will have rights, then it seems to me that we are generally *inter alia* obliged to create the conditions in which these rights can be realised. At any rate the present issue here is not whether such a theory of rights is intellectually acceptable—Attfield may think that it is too weak to be a rights claim—but whether people who accept the rights approach themselves *believe* that their theories of rights generates obligations in respect to people not yet living—and it is clear that many people do believe this.

Likewise with regard to contract theories. I confess I once argued precisely for the problem that contract theories had in accounting for obligations to future generations (since they are not around to be parties to current contracts) (Dower 1983), but in fact a lot depends on just how the contract conditions are specified. Again we need to note that, whatever intellectual difficulties we may have about an ethical theory, if those difficulties are not felt by their proponents and they think that they have a sound basis of supporting a basic principle, then surely we have to say "so be it".

A similar point can be made about ethical theories that do not include either of the two features of Attfield's thinking—the independent value of nonhuman life and the moral relevance of distant future generations. I happen to agree with him that an adequate ethical theory needs to acknowledge these two dimensions. But I am also aware that a lot of environmentalists may be either enlightened anthropocentrists or hold the view on future generations that we are only required to consider the next two or three generations—and such views can generate acceptance of a principle which whilst not exactly the same as Attfield's would generate much the same powerful practical commitments.

(As we noted earlier, Attfield actually argues that it is only if one accepts biocentric values that certain measures to protect the environment will be seen as important, for example, protecting a wilderness not seen as crucial to human well-being (Attfield 2008). A similar point can be made about accepting obligations to distant future generations of humans, for instance over the long-term hazards of nuclear waste. However, my response is that: first, there may still be a basically similar principle accepted—equal entitlement to the atmospheric commons—only a difference in scope of the domain of the beneficiaries is accepted as the basis for working out the entitlement, and second, since it is only in a small proportion of situations that the consideration of nonhuman biota and distant generations leads to either more being demanded or conflicting things being demanded, we should focus on—and celebrate—the vast areas of common agreement on what is required—in contrast to the standard "business as usual" approach of standard assumptions or the acceptance of the adequacy of relatively superficial change.)

At any rate it certainly would be helpful to getting widespread agreement on such ethical principles if we could accommodate a wider range of ethical theories than Attfield appears to allow, if the programme of serious change is to be a realistic possibility (but this is not say that all ethical theories would support it—far from it).

"Polluter Pays": ethical principle or pragmatic principle?

Another twist to this is how we interpret the "polluter pays" principle. In many ways this hovers in an intellectual space between being an ethical principle in its own right and a pragmatic principle. If it is seen as a practical principle—the general following of which would lead to the right agents generally accepting their obligation to act, and thus the conditions are achieved in which the "fair share of the atmospheric commons" is reached—then it belongs to the latter discussion of practical principles which, alongside others, needs to be included because they are widely accepted, and it will make a serious contribution to solving some of the problems. If on the other hand it is seen as a more basic ethical principle, then it is in a sense a principle which either supplements or replaces the "equal share of the atmospheric commons" principle. As an ethical principle it is not simply an aspect of the latter. The latter is, as Attfield has pointed out, not a historical principle, but the "polluter pays" principle is in a sense a historical principle (in Nozick's sense: Nozick 1974) because it attributes moral responsibility for past actions at least as far as culpable contribution to bad effects is concerned. My point here is not however to argue that this principle ought be regarded as a separate ethical consideration, but that as matter of fact it is a principle that many people find very persuasive, and insofar as the widespread acceptance of this supports by and large a wide range of climate change mitigation actions, it is to be welcomed.

Variety of Pragmatic Principles

In regard to pragmatic principles and organising concepts, Attfield is surely right that we need to accept different options. The reason why we cannot expect everyone to accept one principle is because "horses are for courses" and different organising principles will appeal to different groups of actors. As Attfield implies, one group of people may prefer an integrated conception such as the GDR approach does, others may find the Contraction and Convergence approach more manageable albeit supplemented with concerns about development goals as well. One may accept one of them but also accept that others are not to be knocked if they indeed grab other thinkers. Attfield notes that the White Paper looks at a number of different approaches of various values; likewise Paul Harris discussed six major principles that are involved in discussion in the EU over serious climate change policies (Harris 2008). Their being of different value needs to be acknowledged, but at the same time we also have to work on the basis that people are motivated by the conceptions which they have, for whatever reasons, come to accept. So in such a case, we have to weave a complex course between arguing against what we see as inadequate conceptions, and at the same time acting to maintain the motivating force in different actors, who may, if criticised too strongly, just give up their efforts rather than convert to what one regards as a better basis.

Interpretations of the two ethical commitments

How should we interpret the commitment to the two principles Attfield identifies concerning climate change and world poverty? Let us take the commitment to development goals first. Clearly if we could overnight get everyone up to a decent standard of living—that is, escape extreme poverty—that would probably have been done ages ago. But it is not simple, as the history of international assistance has shown. In any case the effectiveness of even effective aid is arguably undermined by the way the international economic system operates, by the punitive effects of debt servicing and so on. Getting major changes meets resistances and takes time. But in any case the extent to which commitment to reduce extreme poverty can be harmonized with progressive reduction in greenhouse gases emissions is rendered more difficult and controversial if we candidly face and answer the question: what is necessary to generate poverty reduction on a wide scale (beyond the targeted and limited efforts of NGO agencies)? There are really two difficulties here.

First, unless poorer countries commit themselves to radically egalitarian policies—which they show no signs of wishing to do any more than rich countries have done—then the likely way in which serious poverty could be alleviated is by some general growth in these countries. As part of that general growth economic growth for the very poor will occur either through a trickle-down process or thorough modest redistributive measures. But it is this general growth in poorer countries that is likely to continue to add pressures on the environment especially if it does not involve new green technologies they will not have without extensive technological transfers from the North. This is quite apart from the possible (though in my opinion not inevitable) increased negative impacts of poor people on the environment as they become less poor. (It is not inevitable, since extreme poverty has its own kinds of negative environmental impact.) Unless there is a dramatic and conspicuous move in rich countries radically to reduce their levels of material affluence and associated carbon impacts, poorer countries will not see reason to do so.

Second, in any case, apart from the generally accepted causal connections between poverty reduction and general increase in affluence in poorer countries, many will argue that even if the North radically reduces its levels of material affluence, poor countries which missed out on this have a right to a much better level of material well being, well above the level needed for escaping from the conditions of extreme poverty. This is a serious ethical challenge, since to some extent this is a morally reasonable expectation, and it will not be realised without further environmental damage unless there are appropriate new technologies facilitated by the North. On the other hand, without a serious and explicit questioning of the lifestyles of India's or China's or other poorer-but-not-so-poor countries' well-off people—alongside a serious questioning of the same in rich counties as well—there is little realistic chance of change. I come to that issue shortly.

Turning to the interpretation of the commitment to the equal share in the absorptive capacity of the planet, it is generally accepted that this is a progressive commitment. That is, it is generally held that what we need to do is to achieve a target of cutting our emissions by 80% by 2050. But, as Monbiot points out, it makes a big difference what kind of curve is achieved between now and then—significant cuts earlier on, or the majority of cuts coming later on (Monbiot 2006). Indeed it is clear that if our commitment is to aim at stabilising greenhouse gases at c. 2 degrees above pre-industrial levels (the consequences of which appear bad enough anyway), then more cuts will be needed later, if the reductions in the curve take place later.

What factors determine what governments do about this? On the one hand, there is the international compliance issue—if other countries generally make significant changes it is easier for any given country to make the same kinds of changes—easier both in the factual sense of it being more likely and also in a normative sense that governments will feel—and maybe justifiably—that they are not required to put their own country's citizens at an undue disadvantage by accepting more than what other countries are doing. On the other hand, there is the democratic issue of what electorates within countries are prepared to authorise. What electorates are prepared to authorise depends on what their own priorities are, over issues liked penalties, incentives, rationing, or voluntary life-style changes, and how they understand their own responsibilities. This is the main focus of the rest of the chapter.

Mediated responsibilities

In the third paper that I mentioned, on mediated responsibility, Attfield identifies a number of ways in which our responsibilities are mediated, and in at least two important respects this applies to climate change ethics (Attfield 2009). One is the sense in which one's actions are part of a set of actions the cumulative impact of which is seriously negative: it is not so much the effect of one's own individual acts as the effects of the class of acts of which they are instances: Attfield quotes Parfit in this regard (Parfit 1984: 70; 78-82), but in fact the idea had already been elegantly expressed by J. S. Mill when he said that the wrongness of an action was in virtue of its belonging to a class of acts which are generally injurious to society (Mill 1910 [1859]): Mill's example was lying, but it equally applies to our frequent acts of fossil fuel emission. It is an important part of what I have called "the ethic of unintended consequences" (Dower 2007). The other form of mediated action is where, whilst we do not directly do some damage, other people do some damage in actions they would not perform but for the fact that we perform the action we do. Most of the impacts on the environment we are responsible for are cases of other people doing things so that we can do the things we wish to do, whether, for instance, it is getting electricity for our gadgets, or having food and other goods transported from far away for us to consume or use. These are cases of the ethical significance of what is done on our behalf to enable us to do, have or be what we want to do, have or be. These senses of mediated responsibility as applied to climate change clearly show how each individual is implicated in his or her general behaviour in climate change issues and suggest a robust account of what is required of each individual as an agent, to which I turn shortly.

In the second half of the paper he turns to the problem of politics: he argues rightly that countries also have mediated responsibilities, over their direct contributions to climate change or their being part of causal chains that involve other countries elsewhere in the world doing what induces climate change, but notes that the problem of getting countries to take their responsibility seriously is tied up with the fact that countries are, especially if they are democracies, meant to do their citizens' bidding. The democratic issue is merely one important aspect of a more general problem that in most countries governments are meant constitutionally to give priority to their current nationals, and this goes against a robust pursuit of global targets based of the long-term interests of all people now and in future generations. What of course we need to encourage is higher levels of cooperation between countries which see that it is in their own interests to take measures to cut back on carbon emissions.

The main thesis of this article seems to be right, but I want to take the line of thought further and link it to the earlier papers in providing a mild corrective to the explicitly state-centric approach of these papers and indeed in most people's discussions of the ethics of climate change.

The democratic issue

Let us take the second issue first. How do we get round the following democratic paradox? From a global or cosmopolitan point of view, the idea that political communities—large of small—should be governed democratically seems a sensible thing to claim, but it has of course the consequence that in any given democratic polity if the majority clearly don't want their governments to pursue a cosmopolitan policy, then in one sense they ought not do so, even if from the point of view of an individual citizen who is cosmopolitan they clearly ought to, for instance on climate change policies. The solution as I see it lies in the increasing acceptance in electorates of the global or cosmopolitan point of view (on climate change, world poverty issues etc). Now the truth is that if governments are going to make really significant and perhaps painful steps to contribute to new policies (and maybe ahead of international consensus), then this will require electorates that are well informed and generally persuaded that this is what their governments ought to be pursuing and that they are therefore willing to accept the consequences of tough policies—tax incentives, rationing or whatever.

This I want to suggest is not likely to happen unless individuals are in significant numbers willing to make judgements about their own responsibilities as moral agents to contribute to the process of changing the public culture *vis à vis* climate change. We need clearly to distinguish between conforming to laws and moral norms and acting as ethical vanguards. That is, we need to distinguish between the idea of agents conforming to new laws or to new socially sanctioned mores about acceptable carbon emissions behaviour—and perhaps a lot of people would now be willing to conform to laws and mores *re* climate change behaviour IF they become well established—and the idea of agents who are willing to put significant amounts or time, effort and money into campaigning for these changes and/or making personal life-style decisions to cut back on their carbon footprint voluntarily and well ahead of the established or prevailing norms of behaviour.

The moral commitment of individuals

There are two points to be made about the latter kinds of moral commitment. First, I contend, without these kinds of vanguard actions by significant minorities, the wider changes in law and social mores will not happen, or will not happen on the scale and with the pace that is required. This is of course an empirical claim, and all I can say is that I think it is plausible. But second, and more significantly, there is the question of how we assess such behaviour. We may commend it but do we see it as in a sense supererogatory—something some people may do but well beyond the call of duty—or should we rather see it as some kind of duty which, in principle, applies to anyone, though most people do not accept it? My own view is that we should regard it as the latter, because as a matter of fact it is really what we all ought to be doing if we have much chance of achieving the long-term changes we have in mind. Let us be clear: if this is the case, we do not have to draw the conclusion that most people are to be blamed because they do relatively little to change their ways. We may indeed blame people whose life-styles are clearly carbon-profligate, but it is counter-productive to blame those who are "on board" to some extent.

In short, my own view is that the ethics of climate change should not merely focus on what countries ought to achieve but also on what individuals ought to do, independent of and prior to any legal compulsion or social sanction so to do. This is for two reasons: first, without this focus and a consequent change in what individuals do and believe, the necessary changes in countries' policies will not be enough; second, it is the cumulative impact of individuals' behaviour that will make the difference. This applies as much to the behaviour of rich individuals in developing countries as it does to rich individuals in developed countries. One of the merits of the Greenhouse Development Rights approach is that it expects contributions from everyone above a certain level of material development. This applies of course to vast numbers of people like most of us here who are relatively well-off, but it applies particularly to the very wealthy, not just because there are a large number of rich individuals in rich countries whose lifestyles are carbon-profligate, but also because there are now quite a number of really rich individuals in so-called developing countries—in India and China for instance—whose private jets, to use an extreme example of a whole range of life-style choices—do the same damage as the private jets of the rich anywhere else. The trouble with a preoccupation with countries—important as this is—is that (a) it takes away attention from the fact that there is an awful lot of carbon-profligate behaviour in poor countries as well as rich countries (even if the per capita level is still low) and (b) it averts our attention from the issue of the distribution of wealth and of carbon emission levels within all countries, rich or poor.

The level of commitment for individuals

There is a further issue, if we accept that individuals ought to change their behaviour prior to and independent of law or social sanction: how much and in what ways should we do so? We saw earlier that what governments, it is now argued in the climate change lobby, ought to do is move *progressively* towards 80% reduction of carbon-emissions by 2050. If it is asked why they shouldn't make the reductions *immediately* (which would clearly be much better for the environment if it were generally done), the answer will be that it is not politically realistic: it takes time to make radical changes in an orderly and fair way in a large political community, democratic electorates have to be won over, international compliance is required and so on. This may be so, though we need to note that the rate of possible change is by no means fixed, and that the speed and manner of changes made by governments is partly a function of what other countries' governments also influenced by their electorates do, and also crucially of what their electorates actually want them to do and want them to do because these citizens come to accept new *moral* priorities (in this sense, "can implies ought" not the other way round)—hence my earlier emphasis on changes in individual thinking and action.

But if we turn to the individual, what should we say about him or her? Is it only realistic that I can only cut my carbon emissions by 80% by 2050 (actually for any particular individual it might be anything, such as 50% or 200%)? Hardly. For many people, if they had the will to do so, they could do it tomorrow. It would involve some rather painful decisions—decisions which most of us with lifestyles that are carbon-intensive are not really prepared to take just like that. There are now facilities—e.g. web-sites—where people can do carbon-inventories for their own lifestyles and know how much they have to do to reduce their carbon emissions (including their share of what is done by others to enable them to do what they do) to the level strictly equal to their entitlement to the atmospheric commons as discussed earlier. Of course I know well that it is not as simple as this: as things stand, there are lots of individuals—usually the very poor, some in countries like the UK but most in poorer countries—who, through no choice of their own, are well below their per capita entitlement, so I can argue that I can exceed my entitlement, especially if I am prepared to offset my carbon emissions in various ways—putting money into renewable forests, paying for a solar panel or loft insulation or whatever, or if I do other morally compensating things like helping with third world development. We can also argue—maybe special pleading if you will—that some of what we do is trying to contribute to positive changes (academics going to environment conferences are good at this kind of argument).

All these moral calculations are relevant—at least for making us feel not too bad about ourselves—but I do wonder if they get to the heart of the issue. This is because most of us do not, really in our heart of hearts, believe that something like an entitlement to an equal share of the atmospheric commons acts as a simple generator of a precise duty that applies strictly and immediately at the level of individual action, or that we can only avoid it if we engage in a lot of moral casuistry to get round it. For one thing it is too abstract to have that kind of moral grip; second, it is dependent on a lot of scientific assumptions the parameters of which are likely to change over time; and third, the moral idea here is more of a desideratum or ideal standard we have a duty to move towards, but not strict duty. In a way our response to this is similar to our response to Peter Singer's famous argument that in response to world poverty we ought to give money (and time and energy) to the point of marginal utility. With this likewise we can add qualification after qualification to temper it, but in the end it represents a kind of ideal rather than an ethical duty in the strict sense.

Whether or not we should treat the ethical demands of climate change mitigation as being on a par with the ethical demands of helping alleviate poverty is one of those key questions that need addressing as part of this debate. At any rate on one of the approaches that Attfield favours—Greenhouse Development Rights—the two demands for poverty reduction and climate change mitigation go together and this seems right. Whilst I am inclined to think—and I am still thrashing this over—that the moral demands to address world poverty and to adjust our lifestyles in the face of climate change are not as extensive in either case, as Singer argues or as a certain interpretation of atmospheric commons entitlement might have it, I am clear that in both cases our moral responsibilities are much more demanding than is commonly supposed, and require of us actions that go well beyond what law or social custom dictate, and that the ethics of climate change needs to pay as much attention to this question of what individuals should do *here and now*, as to what targets countries ought to set themselves *for the future*—partly because the earnest with which countries both set targets and then pursue them is largely a function of what lots of individuals—you and me—prioritise now in their own lives. With this "constructive amendment" to his general approach to the ethics of climate change I hope Attfield will be largely happy.

References

Attfield, Robin, 1983, *The Ethics of Environmental Concern* (Oxford: Blackwell).

—. 1994, 'Rehabilitating Nature and Making Nature Habitable', in Attfield Attfield and Andrew Belsey (eds.), *Philosophy and the Natural Environment*. Cambridge: Cambridge University Press.

—. 1999, *The Ethics of the Global Environment* (Edinburgh: Edinburgh University Press).

—. 2003, 'Global Warming, Justice and Future Generations', *Philosophy of Management,* Vol. 3, No. 1: 17-23.

—. 2008, 'Climate Change: the ethical dimension', conference paper, Padua.

—. 2009, 'Mediated Responsibilities, Global Warming and the Scope of Ethics', *Journal of Social Philosophy*, Vol. 40, No. 2: 225-236.

Baier, P., T. Athanasiou, and S. Kartha, 2007, *The Right to Development in a Climate Constrained World: the Greenhouse Development Rights Framework.*

EcoEquity: www.ecoequity.org/docs/TheGDRsFramework.pdf

Brown, D. et al., 2006, *White Paper on the Ethical Dimensions of Climate Change* (Philadelphia: Rock Ethics Institute).

Dower, N., 1983, 'Ethics and Environmental Futures'. *International Journal of Environmental Studies*, Vol. 21: 29-44.

—. 2005, 'The Nature and Scope of Global Ethics and the Relevance of the Earth Charter', *Journal of Global Ethics*, Vol. 1, No. 1.

—. 2007, *World Ethics—the New Agenda*, 2nd edn. (Edinburgh: Edinburgh University Press).

Elliott, R., 1992, 'Faking Nature', *Inquiry*, Vol. 25: 81-93.

Harris, P., 2008, 'Implementing Climate Equity: the case of Europe', *Journal of Global Ethics*, Vol. 4, No. 2: 121-140.

Harris, P., 2009, *World Ethics and Climate Change* (Edinburgh: Edinburgh University Press).

Meyer, A., 2005, *Contraction and Convergence: The Global Solution to Climate Change*, Schumacher Briefing No. 5. (Totnes, Devon: Green Books).

Mill, John Stuart, *Utilitarianism* (1859), in Mill, 1910, *Utilitarianism, Liberty and Representative Government* (London: J.M. Dent & Sons (Everyman)).

Monbiot, G., 2006, *Heat: How to Stop the Planet Burning* (London: Allen Lane).

Nozick, R., 1974, *Anarchy, State and Utopia* (Oxford: Blackwell).

Parfit, D., 1984, *Reasons and Persons* (Oxford: Clarendon Press).

Rawls, J., 1971, *A Theory Justice* (Oxford: Oxford University Press).

Shue, H., 1994, 'Equity in an International Agreement on Climate Change', in *Proceedings of IPPC Workshop*, Nairobi, July.

Singer, P., 2002, *One World: The Ethics of Globalization* (New Haven, CT & London: Yale University Press).

Chapter Three

Reply to Nigel Dower

Robin Attfield

Nigel Dower has contributed greatly over a long period to the ethics of development and of the environment. Largely I agree with the various points made in his chapter, and in particular the point that individuals have responsibilities in environmental matters. In several places he has taken my thinking forward to a further stage. Some minor qualifications, however, are in place.

(1.) I am not sure that Parfit's point about some actions being wrong because of the set they belong to had already been made by Mill. Mill's point was about classes of action, such as acts of linguistic deception, and the difference made to society if people's confidence in not being deceived were to be undermined (Mill, 1910, 21). Parfit's point was about sets of related actions with imperceptibly small impacts that cumulatively make a large difference (Parfit, 1984, 70, 78-82) such as (we might say, although he did not) acts of emitting carbon dioxide through air journeys, which jointly change the proportion of carbon dioxide in the atmosphere. There are certainly parallels, since Mill's point is about impacts on the cultural atmosphere, which might be thought analogous to impacts on the physical atmosphere. Yet Mill was surely writing rather about how the shared trust of members of society can be eroded by a number of acts of deception that would otherwise do little harm; whereas Parfit's point, or at least the way in which I was using it in the 'Mediated Responsibilities' paper (Attfield, 2009), concerned sets of acts each of which cumulatively contributes to the same serious atmospheric change. Mill's point supplies a ground for rule-consequentialism, while Parfit's works in a different way, urging that the consequences of token acts and omissions must be considered in the light of wide (and often world-wide) contexts, including the actions of others (what we might call a consequentialist version of Kantian ethics). To this extent, Parfit was, I suggest, saying something new and original, although I can see that there was much in common between Mill's move and his.

(2.) While I agree that alliances are needed with people whose ethics are based on rights, and there again with those who derive everything in ethics from contractarianism, and while I try to practice this kind of alliance-building in practical debates (including some that are ongoing), I am less sanguine than Dower about rights theorists having something coherent to say about the rights of future people. I fully agree that we should be concerned about the quality of life of future people, but do not regard the relevant obligations as owed *to* those people, or as corresponding to rights that they have (or even will have) against us.

This is for Parfitean reasons. Setting aside the people of the future who have already been born or conceived, the rest consist of multiple alternative populations, which cannot all exist together; their identity is not only unknown, but has yet to be determined by the parents who eventually bring into being some and not others. So is it possible for all these possible people to have rights, for example against the current generation, for example to be brought into being? If this were the case, the current generation would turn out to carry contradictory obligations to bring into being pairs or trios etc of alternative occupants of the same womb or of the same living space, because of the rights that each member of these pairs or trios would have. (And if we instead say that the rights belong not to possible people but to actual future people, the problem becomes that we can have no idea which these will be, or therefore which rights we are called on to honour, at least in matters of bringing them into being.)

Parfit explains this in a different way; rights can only be carried by those who can have them honoured or disregarded, and who can thus fare better or worse in different worlds. But many possible people exist only in one possible world. So, until some of them actually come into being, we cannot coherently speak of their rights. What we can do is try to bring it about that whichever of the possible people come into being have as high a quality of life as possible. We should do this because it would be wrong not to do it, but not because we owe it to future people who are at present merely possible people, nor because of their rights.

But this said, those current people who regard future people as equally capable of carrying rights as our contemporaries may well be motivated by these beliefs into effort to perform their responsibilities, and their assistance may well be required by those of us who recognise responsibilities of similar content, but on a different ethical basis. There is a similar story to tell about contractarianism, but I will set it aside for present purposes.

I would like to add the parallel suggestion that alliances are also needed among *environmentalists* of divergent metaphysical persuasions, as well as among those promoting opportunities for meaningful lives. If so, then possibly people such as Alan Holland should welcome alliances with environmentalist believers in creation, such as Christopher Southgate and probably Nigel Dower as well as I might count ourselves as being, despite finding their arguments unconvincing and their stances unstable, defective or even positively inconsistent; while our metaphysics should arguably be monistic, pragmatic factors surely require applied pluralism in practice.

(3.) Next, I would like to insert a short note on history. Dower is right that, faced with a choice between historical approaches to environmental responsibility and ones that work from the present situation, I choose the latter. There could be a choice between ascribing the burden of paying for mitigation and adaptation to the historical polluters of the industrial revolution (on the one hand) and (on the other) to those who combine being current polluters and having sufficient current resources to contribute to these costs. One reason why these principles diverge is that some historical polluters may not be able in the present to pay for an amount of mitigation and adaptation proportionate to their historical contribution to pollution; Russia may well be an example, and Kazakhstan another. But concern for sustainability and for mitigating climate change seems to dictate that we select principles that can produce the desired effects, and this strongly indicates selecting principles based on contemporary capacities rather than on history.

However, this does not mean that I am committed to rejecting the relevance of historical factors in general, even though I am a consequentialist. Consequentialists can, for example, support rules of rewarding and punishing, both of them backward-looking institutions, because of the positive difference that they make to current and future society, or, in other words, for forward-looking reasons, and can for parallel reasons uphold the desirability of promises (including ones from the past) being kept. (See further my response to Clare Palmer (in this volume).) They can also take into account the historical associations of (say) a landmark as a reason for preserving it, as long as current people are likely to care about such historical significance. I believe that this openness to historical factors is compatible with rejection of history-based stances about justice such as that of Robert Nozick. Some philosophers, such as Holland, are rather keen on taking historical factors into account, and this has made me eager to explain that my overall stance does not oblige me to dissent.

(4.) Finally I would like to add a comment on individual responsibilities. Dower asks whether vanguard actions and behaving in a green or environmentalist manner are supererogatory or obligatory. I agree with him that in some cases such actions are obligatory, and at the same time that taking this view does not commit us to blaming those who do not perform such actions. There is a difference between what it would be wrong not to do, and what people can be expected to do and blamed for neglecting to do.

This said, I think it is worth pointing out that some vanguard actions could still be supererogatory, deeds which are morally desirable without being obligatory. I have written elsewhere about these distinctions, and possible criteria for their general application, for example in *Value, Obligation and Meta-Ethics* (Attfield, 1995), and do not mean to launch into that field here. But let us look at the matter like this. Most if not all relevant vanguard actions and omissions can be regarded as virtuous ones, displaying green traits of character (such as those commended in a recent article in *Utilitas* by Dale Jamieson (Jamieson, 2007)). But far from all virtuous deeds and virtuous abstentions are obligatory. They are deeds that we ought to do (in the sense of actions that we ought, ideally, to do), but not always deeds that it would be wrong not to do. Almost everyone who believes that there is a category of acts that are supererogatory (morally desirable but not obligatory) would also take the view that being virtuous is not always mandatory or obligatory, and is therefore sometimes supererogatory, whether it is a matter of being saintly or heroic (to cite the phrases used by J.O. Urmson when writing about this matter (Urmson, 1958)), or of just doing the decent thing, or even of just being nice and behaving accordingly.

If so, we should not represent all vanguard actions as obligations; indeed to hold this view might take a lot of the fun from vanguard actions, which can be manifested in all kinds of ingenious and eccentric ways, without behaving in ingenious or eccentric ways in these connections becoming obligatory. My suggestion, then, is that while some green acts and omissions are obligatory, such as limiting one's air travel, perhaps, others, including some life-style choices, are not. Examples might include, for example, choices of what to wear. Somewhere there is a cut-off point, where vanguard actions stop being supererogatory and become obligatory; while I have written about the upper limits of duty elsewhere (Attfield, 1995), I leave to the reader the question of where, in terms of green behaviour, that cut-off point is.

So my response to Dower's final suggestion of a "constructive amendment" is to accept it subject to the further "constructive amendment" just presented, and also subject to the further amendment, mentioned at the July 2009 conference by Jenneth Parker, that many of the relevant obligations belong neither to individuals nor to governments, but to companies and corporations, which should be held accountable not only legally but also morally much more often than they are.

References

Attfield, Robin, 1995, *Value, Obligation and Meta-Ethics* (Amsterdam and Atlanta, GA: Éditions Rodopi).

—. 2009, 'Mediated Responsibilities, Global Warming and the Scope of Ethics', *Journal of Social Philosophy*, 40.2, 225-236; reprinted in revised form in Ruth Irwin (ed.), 2010, *Climate Change and Philosophy* (London: Continuum), 183-216.

Jamieson, Dale, 2007, 'Why Utilitarians Should Be Virtue Theorists', *Utilitas*, 19.2, 160-183.

Mill, John Stuart, *Utilitarianism* (1859), in Mill, 1910, *Utilitarianism, Liberty and Representative Government* (London: J.M. Dent & Sons (Everyman)).

Parfit, Derek, 1984, *Reasons and Persons* (Oxford: Clarendon Press).

Urmson, J.O., 1958, 'Saints and Heroes', in A.I. Melden (ed.), *Essays in Moral Philosophy* (Seattle: University of Washington Press).

CHAPTER FOUR

CREATION AND VALUE FROM A DARWINIAN PERSPECTIVE

ALAN HOLLAND

Introduction

In his *Life of Zeno*, Diogenes Laertius records that the Stoics divided philosophy into three component parts: Logic, Ethics and Physics (DL 1925, Vol. II, sec. 40). At the same time, using the similes of an animal (composed of bones, flesh and soul), an egg (composed of shell, white and yoke) and a garden (composed of fence, crop and soil) they held that none of these parts could be separated out from one another, but were "mixed together".

Throughout his long and distinguished career, Robin Attfield has both recognised and demonstrated this interconnectedness of philosophy—the fact that any position taken will have ramifications elsewhere. And nowhere more so than in his recent book *Creation, Evolution and Meaning* (hereafter Attfield 2006), in which these connections are teased out with great subtlety, particularly regarding the interplay between physical (including metaphysical) and ethical positions. The result is an imposing and ambitious structure—a fitting subject, therefore, to take as the focus of my "critical apprcciation".

Another feature of Attfield's philosophising is his willingness to engage with issues that matter fundamentally to how we live our lives, whether this be the formulation of a cosmological vision, or the question of how we find meaning in our working lives (e.g. Attfield 1984). One undertakes any critical appreciation of his work therefore with the genuine prospect of instruction and, it may be, of some vertiginous realisation that will require a shift in perspective, a prospect that marks one of the perennial enticements of philosophy. It is in this spirit that I offer the following remarks, which constitute an attempt to identify key points at which the argument of *Creation, Evolution and Meaning* fails to carry me along, and then to get to grips with, and understand, our points of disagreement. The claims to be examined are as follows:

(i) that the cosmological argument can be re-worked into a form that supplies strong grounds for believing in a creator who is "eternal, independent, uncreatable and indestructible" (Attfield 2006, 94);
(ii) that belief in creation and belief in evolution—specifically, evolution by natural selection—are compatible:

> the interim conclusions are that Darwinism and theism are compatible and co-tenable, that Darwinism does not undermine all the varieties of the Design Argument, let alone the grounds for theism, and that good grounds remain for endorsing theism as well as Darwinism (Attfield 2006, 121);

(iii) that the design argument may be re-worked into a form that offers real corroboration for the hypothesis of a (beneficent) creator—this is "the argument from value";
(iv) hence, that evolution may be understood as purposive:

> a reconstructed argument to the purposiveness of evolution can open, therefore, with all the varieties of intrinsic value manifest in the actual world… Since these are just the states that a loving creator, desirous of what is intrinsically valuable or desirable, could be expected to seek to generate, there is good reason to regard evolution as having such desirable, complex states as an end or goal (Attfield 2006, 162);

(v) and finally, that a fully meaningful life is likely to be one enhanced by a cosmic vision such as when we understand nature to be the gift of a creator and understand ourselves in the role of stewards of the planetary biosphere and thus of the fruits of creation:

> for those who endorse belief in creation… [cosmic gratitude] can be recognised as due to the creator, whose gifts and generation of value form the framework that gives life its meaning (Attfield 2006, 205).

(Expressed in Stoic terms, a soil that is not enriched with a religious vision may yield little in the way of fruit.)

Part I

For his re-working of the cosmological argument, Attfield uses the Principle of Sufficient Reason to argue that the existence of material objects requires explanation, since this is a state of affairs that could have been otherwise. Further, since nothing in the material world is capable of providing such an explanation, there has to be an explanation of the material world which lies beyond natural agents and agencies: hence, "Nothing less than a creator will do" (Attfield 2006, 95). The existence of such a creator on the other hand, is not subject to the same requirement, since it makes no sense to ask for a reason for the existence of an uncreatable being. The argument involves an appeal to Nozick's elucidation of the Principle of Sufficient Reason, which in Attfield's formulation states that "there is a sufficient reason for whatever could have been otherwise, except where there is a sufficient reason to the contrary" (Attfield 2006, 57). A subsidiary argument is that in the absence of a creator, the probability of a life-permitting universe is zero or at best "metaphysically improbable" (Attfield 2006, 107-8).

Now of course there will be explanations for why some particular material objects exist rather than others, but it is not clear that we need to step outside the realm of ordinary preceding material circumstances for an explanation in each particular case. But the question at issue here is whether some explanation is required for why there are material objects at all: "for the claim that God is the creator of the world is introduced, in part, to explain there being material things rather than there being none" (Attfield 2006, 57). In effect, then, the creator is introduced, in part, to explain why there is something rather than nothing.

For present purposes, we shall put on one side questions about whether the posited explanation is satisfactory, and how an immaterial being is supposed to bring material things into existence, in order to focus on the prior question of whether an explanation is called for in the first place. Attfield argues that it is, appealing first to the Principle of Sufficient Reason as articulated by Nozick, and second to the unquestionable fact that the existence of material things is clearly a state of affairs that could have been otherwise, a fact which generates the reasonable expectation that there must be some explanation for why things are as they are.

However, before accepting that this state of affairs must have some explanation, one is bound to ask, in light of Nozick's formulation, whether there truly is no sufficient reason for supposing the contrary. And contrary to what Attfield appears to argue (Attfield 2006, 58, cf. also, 95) it is unclear that there can be an explanation for why there is something rather than nothing, or—if there were nothing—for why there was nothing rather than something. Or, to put it in Nozick's terms: despite the fact that, whether there was something or nothing, this situation could have been otherwise (there could have been nothing or there could have been something), there does appear prima facie to be sufficient reason for supposing the contrary. The sufficient reason to the contrary might be formulated as follows:

a) either there is something or there is nothing;
b) either there being something or there being nothing is self-explanatory, or it is explained by something other than itself;
c) therefore, if, as Attfield appears to hold, there being something or there being nothing is not self-explanatory, the only thing available to explain there being something is there being nothing, and the only thing available to explain there being nothing is there being something;
d) but there being nothing cannot explain there being something and there being something cannot explain there being nothing;
e) (d) appears to constitute sufficient reason contrary to the supposition that there must be an explanation for there being something or there being nothing;
f) accordingly, there is no reason for supposing, and every reason for not supposing, that the existence of material objects, per se, requires explanation in terms of something other than themselves.

A second consideration that appears to support the first is this. Attfield contends that even where the Principle of Sufficient Reason is qualified to exclude cases where "there is sufficient reason for there being no sufficient reason", the principle "remains strong enough to cover sets… whose members are existing beings which can be caused to exist" (Attfield 2006, 94). But then we must note that if there were a cause of the existence of the set whose members are existing beings which can be caused to exist, then that set would in turn be an existing being that can be caused to exist. In that event, the question arises whether the set would be a member of itself. If we say that it is, then its cause would be a cause of the member of the set rather than of the set. Hence the class "cause of the set" would remain empty, however often the procedure was repeated. If we say that it is not, then we generate a new set that in turn requires a new cause, and in doing so generate a never-ending requirement.

What the above arguments appear to open up is a crucial distinction between a situation that could have *been* otherwise, and a situation that could have been *caused* to be otherwise. For if such a distinction can be sustained, then we can see how a failure to observe it might make the demand for an explanation of why there is something rather than nothing seem reasonable. However, to repeat the point in a somewhat more formal way, from "it is possible that not-p", there is no compelling step that I can see to "it is possible that, for some X, X brings it about that not-p".

Of course, even if the above reasoning is sound, it would not amount to a disproof of the creation hypothesis. For Attfield could insist that an immaterial being is an alternative to nothing, as a possible explanation for the existence of material things. However, I believe this would be a departure from the spirit of the argument. For if I understand Attfield's position correctly, I believe he is claiming that if we follow the logic of *ordinary* explanation through, we are obliged to recognise that an extraordinary explanation obtains in this one special case—when we are trying, as we must, to explain the very existence of material things. Thus, the possibility of an extraordinary explanation—the explaining of the existence of material things by appeal to an immaterial thing—is the conclusion to the argument, and cannot be introduced as a premise. My simple reply is that if we follow the logic of ordinary explanation through, what we appear to find is that there cannot be an explanation for the very existence of material things.

As regards the subsidiary argument—that in the absence of a creator, the probability of a life-permitting universe is zero or at best "metaphysically improbable"—just two thoughts must suffice. Both involve setting aside the claim, which has been contested above, that in the absence of a creator the probability of a life-permitting universe, or indeed any universe, is zero. One is that, if there is a universe at all, then any possible state of that universe, including one that permits life, is equally improbable as any other, given the complexity of the processes that will have brought it about. The other is that to talk in unqualified fashion of the "probability of life" is as problematic as talking of the "probability of the eye", and for similar reasons. For as Richard Dawkins has ably demonstrated (1988, ch.4), the question is not how an eye might arise from what is not an eye, but how it might arise from what is almost an eye. And this, evolution by natural selection might plausibly hope to explain. Just so, the question about life is not how it might arise from non-life, but how it might arise from what is almost life.

Part II

Attfield is in good company when he claims that belief in a creator and belief in evolution are compatible, for as he himself notes (Attfield 2006,

86), Darwin says much the same thing. Furthermore, Darwin's drift towards agnosticism in later life seems to have been unconnected with his long-held belief in evolution by natural selection (as Attfield again notes; 2006, 87). At the same time, Darwin does remark, in a letter written to a German student towards the end of his life, that "Science has nothing to do with Christ, except in so far as the habit of scientific research makes a man cautious in the admitting of evidence" (F. Darwin 1902, 57). What this suggests is that we need to mark a distinction between the content of Darwinism (or more relevantly perhaps "neo-Darwinism")—essentially the belief in evolution by natural selection—and Darwinism as the expression of a particular habit of scientific research that "makes a man cautious in the admitting of evidence". We might label this latter "methodological Darwinism".

With regard to Darwinism in the former sense, there is little difficulty in admitting the compatibility, but with regard to Darwinism in the latter sense, matters are a little more complicated. For if belief in a creator implies the availability of evidence, the question turns on whether there is indeed evidence for a creator. Hence the question of the compatibility of the two positions cannot be answered as an independent question in its own right, but rests upon the outcome of other issues that are the subject of this paper, such as whether the re-workings of the cosmological and design arguments that Attfield proposes are successful. So far as Darwin himself is concerned it has to be said that in the latter part of his life, at any rate, he rarely wavers from the view that "the whole subject is beyond the scope of man's intellect" (*ibid,* 57).

From these reflections, and pending the outcome of other discussions, one has to conclude that there is some doubt as to whether Darwinism in the latter sense is compatible with belief in a creator, since this implies the admitting of evidence, though it may well be compatible with some degree of faith in a creator, prompted perhaps by "the deep inward conviction and feelings which are experienced by most persons", and to which Darwin readily admits that he is susceptible (*ibid*, 60). At the same time, he makes it abundantly clear that he "cannot see that such inward convictions and feelings are of any weight as evidence of what really exists" (*ibid*, 61).

Part III

A major highlight of *Creation, Evolution and Meaning* is its bold reconstruction of the design argument, conducted in chapter 8 and built upon Keith Ward's "argument from value" (Ward 1982, 89-120). According to this reconstruction,

> design on the part of a purposive and loving creator is a reasonable interpretation of the immense amount of value in the world as we find it (Attfield 2006, 151).

This statement throws into relief the central importance to the argument at this point of the concept of value. In doing so, it also throws into relief a potential point of vulnerability in the argument. For in the first place, the argument assumes an objectivist understanding of value judgements. In the second place, the argument assumes that we can form an estimate of "the amount of value in the world". And both assumptions may be questioned. We shall discuss each in turn.

First, it seems clear that both the argument from value, and the argument to the purposiveness of nature which depends on it, do indeed presuppose the adoption of an objectivist account of value. For only on this assumption will the argument have unconditional force. On a perspectival view of value judgements, for example, the argument from value to God will become dependent on the value perspective that is adopted, as will the argument to the purposiveness of nature. Hence the arguments will have only conditional force. Now Attfield himself will certainly agree with this assessment, and is more than ready for it. For he has long been a stout defender of the objectivist account, rejecting all others as in various ways inadequate; and now is not the occasion to rehearse all the possible arguments and objections that bear on such a major topic. But it is, perhaps, an occasion to set down the reasons why I, for one, have (so far) been unable to concur with his stance on this matter.

As a preliminary point we should be clear that our discussion primarily concerns ascriptions of intrinsic value expressed by means of "value-judgements" (these must be distinguished from statements of the form "A is of value to X", which may for the most part be construed as empirical claims assessable in the normal way as objectively either true or false). Along with his stout defence of the objectivity of value judgements, Attfield has equally stoutly resisted perspectivism; but only, it seems to me, in a rather casual formulation of that position—as the view that what is valuable is what is valued from a certain perspective. To this he rightly objects that something can be valuable without being valued at all. For intrinsic value is to be defined in terms of "what there is reason to desire, cherish or foster in virtue of the nature of the state or object concerned" (Attfield 2001, 152; see also Attfield 2006, 154), and what there is reason to desire, cherish or foster may be something that no one ever has valued, does value or ever will value. However, it is not clear why perspectivism can not be understood, rather, as the view that there may be different perspectives on what there is reason to cherish. Thus, Eric Katz (1985,

253, n.28) thinks there is reason to cherish the small-pox virus. David Schmidtz (1998, 64) does not. Is there really an objective fact of the matter here? (One might add to this a certain puzzlement as to how a value-judgement can be advanced in the absence of any perspective from which the value judgement is made. Note that the puzzlement concerns value judgements only, not ascriptions of truth in general.)

Clearly, nothing can be settled by rhetorical question, so let us pursue the objectivity path and see where it leads. Initially, it leads to a problem of determinacy. For whilst A might think there is reason to cherish X, B might think there is reason not to cherish X. And these two thoughts are not in contradiction to one another. But if both are construed as objectively true, then the question of whether the small-pox virus, or any other entity concerning which there may be competing reasons, is valuable, becomes indeterminable. Alternatively, one may take the view that one position is wrong. And indeed, one must take this view if one position is described as the view that there is reason to cherish X, while the other is the view that there is no reason to cherish X, since these positions do contradict one another. But then we encounter what seems to me a serious conceptual difficulty.

To see this, consider what might be called "the Caligula problem". The Roman emperor Caligula is reported to have been particularly fond of tearing the wings from flies. One might imagine this being elevated into a national sport. One might even imagine it being declared the be-all and end-all of existence—a life without the activity is just not worth living. If values are objective, in the sense that they hold true no matter what humans may or may not believe, this accords values the status of facts. In that event, and even though no one with the possible exception of Caligula believes such a thing, it could in principle be a fact that tearing the wings from flies is the be-all and end-all of existence. For that is the way with matters of fact. Despite what we all think we have every reason to cherish, in principle the truth could prove otherwise. Since the problem arises from the resolve to separate what there is (objectively) reason to cherish from what anyone, anywhere, might think they have reason to cherish, the only solution to the problem that I can see is to resort to perspectivism. From a perspectivist viewpoint, there could be no such fact, because values, as distinct from facts, are essentially matters of judgement made from some perspective or other, and as such are ever open to critical scrutiny. But a judgement such as Caligula's would not long stand scrutiny. The point is that the human perspective contains the resources with which to critique Caligula's predilection—at least one must fervently hope and believe that it does.

But suppose for the sake of argument that these objections should turn out to be unfounded. Nevertheless, another source of difficulty remains. This concerns what grounds there might be for confidence in the assertion that "there is an immense amount of intrinsic value in the world". On the one hand, it seems to me that Attfield does an excellent job in disposing of many of the objections that have been raised to the belief in a beneficent creator that are based on the undoubted existence of "evil" in the world, both moral and natural (Attfield 2006, ch.7). In particular, he makes a persuasive case for saying that

> we have no reason to believe that a world with a better balance of good over evil than the actual world is possible, or that the actual world is not a world that a good God would create (Attfield 2006, 141).

On the other hand, in support of this claim he cites Reichenbach's observation that the task of showing that a world with a better balance of good over evil is possible "cannot easily be accomplished, and is suited to nothing less than an omniscient mind" (cf.Reichenbach 1982, 116). Precisely so. But the same surely applies to the task of showing that a world with a better balance of good over evil is not possible. Thus, while Attfield rightly denies the necessity for attempting to demonstrate that this is the best of all possible worlds, the belief in a beneficent creator does require us also to believe both that there is a preponderance of value in the world, and that no greater preponderance could possibly have been brought about. But it strikes me that to provide evidence for both these propositions is a very tall order indeed, and that if we go with W. K. Clifford's admonition always and everywhere to apportion our belief to the evidence (1947, 77), then the only reasonable stance to take is that of agnosticism.

Nothing daunted, Attfield does indeed take on the task of defending both these propositions, but with uncertain success. Reviewing such natural evils as waste, pain, suffering and aspects of parasitism and predation, he convincingly demonstrates that these are necessary features of a system that is to produce all that we find valuable in the world. But this is some way short of demonstrating, or even providing in itself the slightest reason to believe, that there is a preponderance of good over evil in the world, and this, over an indefinite stretch of time. It is difficult to envisage the cosmic calculation that would provide such evidence; nor is it clear that the concept of value can bear the weight that is being put upon it. This is not to deny the possibility of, indeed the practical necessity for, comparative judgements of value: shoes for Janet, or gloves for John, on this month's budget? But these are localized judgements, presupposing limited resources, specific attachments and finite projects; they do not readily translate to the cosmic canvas. The class of valuable things, in all its manifold and various forms, admits of no obvious commensurating unit. And even if we take an apparently simple case—a comparison of the intrinsic value of the life of the cuckoo and the life of mistletoe respectively, which Attfield claims to be able to "include in the reckoning, alongside their various impacts, good and bad" (Attfield 2006, 130)—the nature and basis for such a reckoning remains elusive. Nor is there clear agreement that the phenomenon of suffering should even enter into such a calculation. There are, after all, respected viewpoints which hold that no amount of benefit can justify the suffering of even a single sentient creature.

Part IV

Many have tried to discern a trend in evolution, whether this is towards diversity, merely, as claimed by Aldo Leopold (1949, 216), or more ambitiously towards complexity or even consciousness (Ward 2004). Attfield's ascription of purpose to evolution is both more modest and more subtle. Building on the argument just discussed, the trend is said to be towards the generation of intrinsic value. Indeed, in arguing that the laws of nature themselves may be thought to be designed precisely with the prospect of bringing about a cosmos imbued with value, he is taking a leaf out of Darwin's own book, in particular the *Essay on Theology and Natural Selection*. Here, Darwin suggests that "the laws of transportation were created with reference to successive development" which Attfield glosses (quite plausibly) as the suggestion that "the laws of nature could have been so devised that the various species would develop from one another over the course of time" (Attfield 2006, 84-85) although, as he goes on to note, this was not Darwin's eventual view.

However, for reasons adumbrated elsewhere (Holland 2009, 506-509), a full blown neo-Darwinian understanding of evolution offers distinctly infertile territory for the discerning of trends, whatever their destination. On this reading, what Darwin affords us is the truly amazing—but at the same time, astoundingly humdrum—possibility that the world as we know it came about largely as a result of innumerable and wholly unremarkable everyday events ("largely", because we now believe that several catastrophic events also played their part). Where we are now—the current state of the world—is simply where a purely historical, happenstance process has left us. Thus the only "law" that evolution follows is what John Herschel called "the law of higgledy-piggléty", as recorded by

Darwin in a letter to Sir Charles Lyell (Burkhardt 1996, 208). The conclusion I draw is that a neo-Darwinian belief in the sole operation of natural processes can only ever justify the recognition of a *de facto* trend—that is, a trend for which there is no underlying explanation; it can never justify us in postulating a direction and therefore can never justify us in detecting purposes.

To consolidate this objection a little more formally, let us temporarily discount our earlier objections, and allow that the universe does indeed manifest a preponderance of value. It bears remark that Darwin himself in fact concurs with this view, remarking on "the generally beneficent arrangement of the world"; but interestingly, he attributes this arrangement not to design but to "the effects which we might expect from natural selection" (F. Darwin 1902, 59). Attfield argues that where we have ended up—that is, in a world imbued with value—was where we were likely or even obliged to end up. My objection, leaning on the understanding of Darwin that I have tried to outline, is that in a world of radical contingency there could be no such guarantee. I would argue as follows:

[i] according to neo-Darwinists, evolution is characterised by happenstance and radical contingency;
[ii] the appropriate model of change in such a process is "the random walk"—a stochastic process whereby each next step is determined by pure chance;
[iii] where you end up following a random walk is pure chance;
[iv] it was pure chance that we ended up with a world of value;
[v] therefore, the existence of value cannot have been the result of design.

It should be noted that the appeal to chance in no way implies the abandonment of explanation. For a chance event or situation is precisely one in which everything that contributes to the event or situation is explained, thus leaving no further explanation necessary for the event or situation itself.

Now it has been claimed, for example by Ward and also by Attfield, that the issue turns on what is the most reasonable, or best, explanation for the preponderance of value in the world—the appeal to chance, or the appeal to design. Both Ward and Attfield affirm the latter (Attfield 2006, 163). But if the choice lies between a form of explanation, each step of which is readily intelligible, and a form of explanation that is acknowledged to lie beyond human comprehension, then I must demur.

Part V

Finally, it is an underlying theme of *Creation, Evolution and Meaning* that a fully worthwhile life is most likely to be attained under the hypothesis of a divine and beneficent creator, with all that this implies. I have elsewhere given reasons for questioning this claim (Holland 2009). In sum, they are:

a) that the prior stipulation of a purpose to human existence detracts from, rather than enhances our ability to lead individually purposeful lives;
b) that the prior assurance of a "solution" to the problem of suffering—an assurance that whatever the degree and kind of suffering in the world, it will ultimately be redeemed—not only makes suffering into what it is not, namely a "disvalue" that can in principle be cancelled out, but also obliterates the kind of courage that would be called for to face up to the possibility of irredeemable suffering;
c) a prior guarantee that meaningful lives are possible removes what can be seen as one of the most central and challenging features of the human condition, the fact that the question of whether meaningful lives are possible lies in the balance (*op. cit.* 515-516).

But a further reason emerges in light of our preceding discussion. As previously argued, it would appear that meaningful lives lived under the creator hypothesis do rather presuppose both that there is a preponderance of value in the world, and that no greater preponderance could possibly have been brought about. On this view, then, the possibility of a meaningful life becomes hostage to a value calculus of cosmic proportions that lies well beyond our human capacities. It is therefore a stance upon which we cannot possibly place any reliance. Or so it has been argued. Without the creator hypothesis, on the other hand, and fully acknowledging that worthwhile lives cannot be lived in solipsistic indifference to our cosmic plight, it can plausibly be argued that a sufficiency of value, merely, is the plausible and attainable backdrop for the living of a meaningful life. And given this sufficiency of value, of which we can reasonably be assured, then perhaps worthwhile and meaningful lives can be lived in the merest of circumstances, such as the presence of some other, human or non-human, to love. My conclusion is that far from enhancing the prospects for a meaningful life, the creator hypothesis casts the possibility of such a life into the greatest of doubt.

References

Attfield, Robin, 1984, 'Work and the Human Essence', *Journal of Applied Philosophy*, 1.1: 141-150.

—. 2001, 'Postmodernism, Value and Objectivity', *Environmental Values*, 10.2: 145-162.

—. 2006, *Creation, Evolution and Meaning* (Aldershot: Ashgate Publishing Ltd).

Burkhardt, F., 1996, *Charles Darwin's Letters: A Selection* (Cambridge: Cambridge University Press).

Clifford, W. K., 1947, *The Ethics of Belief and Other Essays* (London: Watts & Co.).

Darwin, Francis, 1902, *Charles Darwin: His Life* (London: John Murray).

Dawkins, Richard, 1988, *The Blind Watchmaker* (London: Penguin Books Ltd).

Hicks, R. D., trans. 1925, *Diogenes Laertius: Lives of Eminent Philosophers*, in 2 volumes (London: Heinemann Ltd).

Holland, Alan, 2009, 'Darwin and the Meaning in Life', *Environmental Values*, 18.4: 503-518.

Katz, Eric, 1985, 'Organism, Community and the Substitution Problem', *Environmental Ethics*, 7: 241-257.

Leopold, Aldo, 1949, *A Sand County Almanac* (Oxford: Oxford University Press).

Reichenbach, Bruce R., 1982, *Evil and a Good God* (New York: Fordham University Press).

Schmidtz, David, 1998, 'Are All Species Equal?', *Journal of Applied Philosophy*, 15: 57-67.

Ward, Keith, 1982, *Rational Theology and the Creativity of God* (Oxford: Basil Blackwell).

—. 2004, 'Theistic Evolution', in *Debating Design: From Darwin to DNA*, ed. William A. Dembski and Michael Ruse (Cambridge: Cambridge University Press), 261-274.

CHAPTER FIVE

REPLY TO ALAN HOLLAND

ROBIN ATTFIELD

Alan Holland, a renowned philosopher for whose work I have the greatest respect, has raised some searching questions and penetrating problems. I consider here his remarks on each of his five themes in turn.

Holland claims in Part I that in effect I introduce God as creator to explain why there is something rather than nothing. But actually, because God, if able to bring about material things, would also be something (and certainly not nothing), the existence of God could not itself possibly explain why there is something rather than nothing. Nor was I suggesting otherwise. Indeed I argued in *Creation, Evolution and Meaning* (Attfield 2006, 94) that there cannot be a sufficient reason for there being contingent positive states, since necessary explanations cannot explain what is contingent, while contingent states, if put forward as explanations, would be among the states to be explained, and yet these are the only possible kinds of explanations.

Unfortunately Holland's mistaken belief that I was seeking to explain why there is something affects his subsequent would-be refutation of my reasoning. For he represents my reasoning as based on the premise that there being something and not nothing is not self-explanatory, and proceeds to argue that we have sufficient reason to hold that there need not be an explanation for there being something (or indeed for there being nothing).

My reasoning, however, relates not to there being something, but to there being a material universe. The Principle of Sufficient Reason, in the form that I endorse, does not require there to be a sufficient reason where (as in the matter of there being contingent positive states or in that of there being something) no such explanation is possible (see the reasoning presented above). But where there is no sufficient reason why an explanation is impossible, it maintains that there is an explanation; and such, I suggest, is the case with the existence of the material universe.

Holland's refutation, with its steps (a) to (e), leaves this reasoning untouched.

However, Holland proceeds to introduce an argument intended to show that the quest for an explanation of sets whose members are existing beings which can be caused to exist is bound to be futile. Yet we know from beforehand that such a quest need not be futile. For the set of my children is a set of existing beings which can be caused to exist, and which clearly has an explanation. The only issue is whether the set of all existing beings that can be caused to exist has an explanation, as the Principle of Sufficient Reason (interpreted as above) claims. The issues raised in Holland's dialectic here were tackled some time ago in William Rowe's article 'Two Criticisms of the Cosmological Argument', in *The Monist* (1970, 441-459). Rowe showed there to my satisfaction that for both finite and infinite sets, the questions of what brings it about that the set has members (rather than none) and has the members that it has (rather than others) can only be answered by reference to something outside the set. This granted, the issue of whether the set of existing beings which can be caused to exist is a member of itself becomes a side-issue. On a material interpretation of sets, the set simply is its members, but we can still ask the same explanatory questions about them, using another sense of "set". On an abstract interpretation of "set", sets with material members are not likely to be members of themselves; but the explanatory questions about such a set having members (etc.) do not go away, even though talk of a set which is not a member of itself happens to be generated when we reflect on sets of this kind.

As for moving from "this could have been otherwise" to "there is a sufficient reason why it is as it is and not otherwise" the missing step is precisely the Principle of Sufficient Reason, with the same proviso as above that this does not apply to states where there is a sufficient reason for there being no sufficient reason. But this kind of explanation is surely an ordinary kind, and not an extraordinary one. In any case Holland has not contrived to show that "there cannot be an explanation for the very existence of material things".

Holland's remaining remarks in Part I relate to the Fine-Tuning Argument (see Attfield 2006, 105-108). And here the issue is not how life might arise from what is almost life, but what explains the stringent conditions in the absence of which there would be no possibility of life being satisfied in the real world. While, under some descriptions, there being life is no more improbable than any other contingency, the universe having constants with values in the narrow bands suited to life when any of the full range of values was equally possible continues to call for an explanation, and thus to supply a contemporary version of the teleological argument to there being a cosmic architect or designer.

This matter is also relevant to the request for evidence of theism that surfaces in Part II. Holland there grants that, where its content is concerned, Darwinism is compatible with theism, but raises doubts about the compatibility of theism and "methodological Darwinism", which involves "the admitting of evidence". Holland illustrates this by referring to a remark of Darwin about Christ and the need for epistemological caution.

Since I too am committed to "the admitting of evidence", I should first register that while the historicity of Jesus seems well established by the evidence of the synoptic gospels, I take seriously also the new evidence from the Gospel of Thomas, and the evidence supplied by Elaine Pagels in *Beyond Belief* that St. John's Gospel was a polemical work in which Thomas is intended to emerge as discredited, while it also makes questionable claims about Jesus' uniqueness and divinity, conflicting with those of the Gospel attributed to Thomas (Pagels 2004). This approach to sifting the evidence about Christ seems to resemble Darwin's.

So I would seem qualified to claim to be a methodological Darwinist, and furthermore that theism and methodological Darwinism really are compatible. Holland suggests that this all depends on whether the re-workings of the cosmological and design arguments that I propose are successful, and implicitly that if not then theism and this kind of Darwinism could be incompatible. But where methodological Darwinism is in question, what is important is the appeal to evidence (as opposed to appeals to feelings, a contrast introduced by Holland at the end of Part II). Yet this is precisely the appeal that I (and many other theists, including Ward) actually make. So even if our arguments somehow fail, our methodological Darwinism seems compatible with theism insofar as our theism at least purportedly rests on evidence in accordance with just such a method.

Before I move on, it should be remarked that when Holland writes of my re-working of the Design Argument in the forms of the Fine-Tuning Argument (Attfield 2006, ch.5), and of my reconstruction of Keith Ward's Argument from Value (Attfield 2006, ch.8), it is less than clear that my re-expression of the traditional Design Argument (in chapter. 5, at pp. 101-106) has come to his attention. This passage is much indebted to the work of Richard Swinburne, and involves an appeal to the evidence of laws of nature, as independent evidence for there being a cosmic designer. This being so, this version of the Design Argument should also be taken into account, if only as evidence that I appeal to evidence. Readers who have hitherto assumed that the Design Argument consists merely in Paley's analogy between a watch and the universe should pause to reflect on the distinctive argument (and thus the evidence) presented in this passage.

In Part III, Holland criticises my revised version of Ward's Argument from Value. His first move is to contest my objectivism about value, offering a perspectivist account instead. He accepts my objections to the kind of perspectivism that claims that what is valuable is what is valued from a certain perspective, but advances in its place the view that "there may be different perspectives on what there is reason to cherish". So far, however, this is an innocuous claim, for there may be different perspectives on most matters (without any implication being entailed that there is no truth of the matter). Further, if two people disagree on whether there is reason to cherish something, they may be appealing to reasons of different kinds (prudential, anthropocentric or biocentric), and this might explain why people disagree about whether there is reason to cherish the smallpox virus. But from this it would not follow at all that the question of whether the smallpox virus should be preserved "becomes indeterminable".

Here Holland introduces the problem of the values of Caligula (who reportedly valued tearing the wings from flies), to illustrate the thesis that values are matters of judgement made from some perspective or other, rather than being facts. Here I suspect that we are at cross-purposes. For while "values" (in the plural) always belong to someone or some society, and are thus perspective-related, this need not be the case with "value" and "valuable", concerning, as they do, what there is reason to desire or foster or cherish. ("Value" in the singular is only sometimes the singular of "values", but is sometimes logically akin to concepts such as "dignity" or "worth", which of course do not have a plural at all.) My view is that the perspective-related nature of "values" in no way indicates that value (as in "having value" or "having intrinsic value") is perspective-related. Indeed in my view it is important that "value" in this sense is not invariably perspective-related (at any rate where there are more than one possible perspectives), since this means that it is possible to compare perspectives in respect of the value of what they approve or commend, and that we do not have to grant that there is no appeal beyond perspectives.

However, if the view is taken that value-ascriptions are relative to a universal perspective which potentially recognises all reasons as reasons, then I have no objection to perspectivism of this kind. Holland's verdict on Caligula's values may supply an example of such a universal perspective, when he writes that "the human perspective contains the resources with which to critique Caligula's predilection".

Holland now turns to criticising my claim that "there is an immense amount of value in the world", and here he accepts the persuasiveness of my case for holding that we have no reason to believe that the actual world is not a world that a good God would create (Attfield 2006, 141). However, he proceeds to suggest that we have no reason either to believe that a world with a better balance of good over evil is not possible, and that I have not persuaded him otherwise. In this connection, he claims that belief in a beneficent creator requires us to believe both that there is a preponderance of value in the world, and that no greater preponderance could possibly have been brought about, propositions which he takes me to set about defending.

But I was not claiming that a world with a better balance of good over evil is not possible. Imagine the actual world supplemented with one extra violet; this sounds like a better balance of good over evil, and a possible one at that. Further, belief in a beneficent creator does not seem to require anyone to believe that the creator could not have brought it about, or is less than beneficent through failing to do so. If so, then theists have no need to deny that such a supplemented world is possible.

However, I did argue that "the system of nature may well be regarded as having an overall balance of value over disvalue" (Attfield 2006, 143). This was concluded in the light of a biocentric account of the value of the flourishing of lives, whether human or non-human, and in the course of a re-working (Attfield 2006, 139-143) of Holmes Rolston's arguments about waste, pain, suffering, predation and parasitism as "necessary features of a system" that produces "all that we find valuable in the world" (Holland's phrasing). Holland indeed allows that I "convincingly demonstrate" that this is the case, but adds that comparisons about the balance of value over disvalue are unreliable, and that the concept of value may not be capable of bearing the weight put upon it here. He also complains that in the absence of any "obvious commensurating unit", comparisons of the intrinsic value of the life of the cuckoo and the life of mistletoe would have an elusive basis. Yet many comparisons are made in the absence of commensurating units, and his problem with this comparison may well relate to his scepticism about the very concept of intrinsic value (which I have defended in an article cited by Holland (Attfield 2001), and have insufficient space to defend again here). But people can and do compare worlds, and whether they are worth preserving or promoting (that is, whether they are sufficiently valuable), not least when they deliberate on how to enhance the actual world, and they are not invariably confused when they do. Nor does it seem far-fetched to suggest that people can compare a world with cuckoos and no mistletoe with a world with mistletoe and no cuckoos (which latter we may be accidentally heading towards unless we preserve habitats better than at present). Inter-world comparisons, then, do not seem to be an insuperable problem, even when they turn on value.

At the end of Part III, Holland refers to "respected viewpoints" holding that "no amount of benefit can justify the suffering of even a single sentient creature". Thomas Hardy held such a view, and Fyodor Dostoyevsky's character Ivan Karamazov seems implicitly to have upheld something similar. Such views, far from being neglected, are considered in *Creation, Evolution and Meaning*, and reasons for dissent are offered at pp. 128 and 137 respectively. In some forms, these views imply that a creator would not be justified in generating any animal life at all if it was going to mean that one sentient creature would suffer a single headache. This, I suggest, further bears out the need for comparisons between perspectives, in which some may defensibly be deselected.

In Part IV, Holland contests my conclusions about evolution having a direction. First, however, he handsomely acknowledges that Darwin may well have held similar views at one stage, and also may have accepted that the world manifests "a preponderance of value". He resists, however, the twin view ascribed to me that "a world imbued with value was where we were likely or even obliged to end up". But I would not assent to "obliged", either in its ordinary sense or in the sense of "predestined"; and because of this, I need not subscribe to the view that the world's being as it is, is somehow guaranteed (a view Holland here ascribes to me).

Holland goes on to deploy his "random walk argument" for the conclusion that the existence of value cannot have been the result of design. However, if either of the arguments for design succeed (whether from the laws of nature or from the world's intrinsic value), then either one or another of the premises of the random walk argument cannot stand or possibly the argument is invalid, maybe through failure to block the possibility that the laws of nature are so structured as to generate value, albeit through random processes. A further possibility is that the opening proposition, which reports the neo-Darwinist view of evolution, supports conclusions for neo-Darwinists (or one sort of neo-Darwinist) only, and that evolutionary theory (or Darwinism as opposed to neo-Darwinism) is open to alternative metaphysical interpretations which, unlike neo-Darwinism, allow value to be generated.

At the end of the section, Holland suggests that explanations of the world's value by Ward and myself are "acknowledged to lie beyond human comprehension"; but I, for one, do not acknowledge this, and have argued in the first four chapters of *Creation, Evolution and Meaning* that the concept of God can be understood sufficiently to allow of arguments to God as creator. Imaginably Holland is here alluding to my endorsement of Reichenbach's view about the difficulty of showing that a better world is possible (Attfield, 141), since that would involve the ability to foresee comprehensively the difference that different laws of nature would make. But it does not follow that human beings cannot follow arguments to the best explanation about the world as it is, or be reasonable in endorsing their conclusion.

In Part V, Holland summarises arguments concerning the relation of meaningful lives and belief in a beneficent creator. His conclusion is that "the creator hypothesis casts the possibility of such a life [sc. a meaningful life] into the greatest of doubt".

There are several problems with these arguments. Why are the "stipulation of a purpose", the "assurance of a 'solution' to the problem of suffering" and the "guarantee that meaningful lives are possible" ascribed to theism all characterised as "prior"? For many theists, there are few or no stipulations, assurances or guarantees, and the actual beliefs held arise out of experience rather than apriorism. Again, why is belief in suffering being redeemable characterised as making suffering what it is not?

Holland now adds a further consideration. Meaningful lives lived under "the creator hypothesis" presuppose that there is a preponderance of value in the world, and that no greater preponderance could have been brought about. (I have contested above the claim that the latter actually is presupposed.) But the possibility of such lives is thus "hostage to a value calculus of cosmic proportions that lies well beyond our human capacities", and thus cannot be relied on. By contrast, meaningful lives lived without this hypothesis depend on nothing more than "a sufficiency of value", "of which we can reasonably be assured", and their possibility is thus less unreliable.

This argument itself presupposes the coherence of the concept of non-relative value recently contested by Holland (Holland 2009). It also asserts that we can "reasonably be assured" that there is a sufficiency of value in life or in the world. But this already suggests that people can reason about how much value there is. However, they need not go much further than this to reach the view that there is a preponderance of value in the world. This is a view with which, as Holland mentions earlier, Darwin concurred in "remarking on 'the generally beneficent arrangement of the world'"; and so it can hardly require "a value calculus … that lies well beyond human capacities". Besides, they have no need to reason about whether or not a greater preponderance could have been brought about; such reasoning is, surely, for metaphysicians rather than for theistic religious believers in general. In the circumstances, the possibility of meaningful lives lived on a basis of belief in a creator ceases to seem far-fetched or inaccessible, particularly as praising God for life's blessings need require no calculation at all.

The possibility of meaningful lives associated with neo-Darwinist beliefs has recently been questioned by Herman Daly (Daly 2002). Difficulties that he raises for meaningful planning of the future turn on what he considers the tendency of neo-Darwinists (a) to reject belief in value and (b) to endorse determinism, and thus accept that there is only one possible future. But it is less than clear that neo-Darwinists need to make either of these moves; Daly seems to have been over-generalising about neo-Darwinism. He does, however, show that there are possible problems for lives lived on this basis being meaningful, even though the problems are avoidable.

Indeed there are probably problems about the possibility of meaningful lives lived on any basis, religious or materialist, and for ones lived on an agnostic basis as well. Sooner than focus on the metaphysical problems closest to our own or others' basis and beliefs, it may be better to focus on opportunities for meaningful work, a topic on which Holland and I are clearly in agreement.

References

Attfield, Robin, 2001, 'Postmodernism, Value and Objectivity', *Environmental Values*, 10.2: 145-162.

—. 2006, *Creation, Evolution and Meaning* (Aldershot and Burlington, VT: Ashgate).

Daly, Herman, 2002, 'Policy, Possibility and Purpose', *WorldViews: Global Religions, Culture, and Ecology*, 6.2: 183-197.

Holland, Alan, 2009, 'Darwin and the Meaning of Life', *Environmental Values*, 18.4: 503-18.

Pagels, Elaine, 2004, *Beyond Belief: The Secret Gospel of Thomas* (New York: Vintage Books).

CHAPTER SIX

PARASITES AND PANGLOSS: ROBIN ATTFIELD ON EVOLUTIONARY THEODICY AND ENVIRONMENTAL ETHICS

CHRISTOPHER SOUTHGATE

My encounters with Robin Attfield have been diverse, even though we had never met until the colloquium in his honour in Cardiff in 2009. I knew of course of his books on environmental ethics, and I was therefore delighted to be sent for review his more recent *Creation, Evolution and Meaning* (Attfield, 2006a). I described this without hesitation as "an important work of philosophical theology", and noted that I was "full of admiration for the clarity of thought that underlies every part of this book" (Southgate, 2008a). I was particularly helped by his treatment of evolution and suffering, which contributed significantly to my own recent monograph *The Groaning of Creation* (Southgate, 2008b).

So I was delighted to receive an affirmative email from Attfield about *Groaning*. It's par for the course in academe that no sooner is one's own book out that another book in the same area appears immediately. A philosophical treatment of the problem of animal suffering was duly published straight after my own book. This was Michael J. Murray's *Nature Red in Tooth and Claw*, which I was able to ask Attfield to review (Murray, 2008; Attfield, 2009a). I will return to Murray's book later in this chapter. But meanwhile I had the extraordinary privilege, at a conference in Crete in 2008, of standing in for Robin Attfield, reading out his paper and fielding questions on the latest version of his argument on stewardship.

So I have had plenty of opportunity recently both to appreciate Attfield's work and even to pretend to be him! What I want to do in this present piece is to comment on the relation between his position on evolutionary theodicy and my own, and then consider how that difference in theological approach might be correlated with a difference in ethical emphasis.

Attfield on Evolutionary Theodicy

One of the great merits of Attfield's treatment of evolutionary suffering is his nuanced and critical approach to evolutionary theory, which is like any good theory full of internal arguments (Attfield, 2006a, 109-14). This bears very much on his final question: "Could things be otherwise?" (Attfield, 2006a, 147-50) and hence on the burden of suffering that the evolutionary theodicist is tasked with addressing. The jury is very much out on the adequacy of natural selection alone as an evolutionary mechanism. Yet the theologian should be very wary of the assumption that God had to input "information" to effect certain key transitions—the implication of Holmes Rolston's position in his Gifford Lectures, published as *Genes, Genesis and God* (Rolston, 1999). I think Attfield may be over-charitable to Rolston, and to Keith Ward, in supposing that they are positing purely naturalistic schemes as explanations for how evolution has given rise to the outcomes that it has (Attfield, 2006a, 149-50). Rolston writes,

> for the key transitions in evolutionary history new information is needed in enormous amounts and… one cannot just let this information float in from nowhere… there is a Ground of Information… otherwise known as God. (Rolston, 1999, 359).

Ward writes in his *God, Chance and Necessity*,

> Taking natural selection alone, it seems to me highly unlikely that rational beings should ever come to exist in a universe like this… I regard evolution by natural selection as a much more insecure and precarious process than seems compatible with the theistic idea of a goal-directed process… a continuing causal activity of God seems the best explanation of the progress towards greater consciousness and intentionality that one sees in the actual course of the evolution of life on earth. (Ward, 1996, 77-78)

In such passages these distinguished authors seem to run the risk of inserting God's influence into gaps in the causal order as described by science, gaps which further scientific understandings are in the habit of closing.

That is not to say that theistic reflections may not inform research programmes in evolutionary science—a good example is Simon Conway Morris's work on convergent evolution, and on the possible yet-to-be-characterised constraints that may exist on what evolutionary forms are ever "tried" (Conway Morris, 2003). But the existence of such constraints, and the reasons underlying them, need to be properly established within the methodology of the relevant sciences, and not grafted in from theology, however rich the connections that might then be established between the resultant science and the theology of creation.

Of the trio of theological sources Attfield addresses at the end of his chapter on "God and evil" I think only Arthur Peacocke truly makes clear the importance of the evolutionary theologian engaging with a purely naturalistic account of the long narrative of evolution (see for example Peacocke, 1990). Any account of the evolutionary process that requires God to make up deficiencies in that process starts to fall into the same traps that beset intelligent design arguments. In particular, it seems inherently problematic to postulate that natural selection, operating on spontaneous variants, can accomplish almost all the evolution of characteristics of organisms, but that just occasionally the process cannot work and had to be either steered or set aside. And, of course, any account that involves any significant element of divine steering of the evolutionary process by efficient causation, as for example in the proposal of Robert J. Russell (Russell, 1998), greatly intensifies the problem of theodicy—it makes God not only the author of a process to which vast quantities of suffering and extinction are intrinsic, but it also posits that God is efficiently active in the process without commuting that disvalue.

That said, Attfield is quite right that evolution does not and cannot establish materialism (Attfield, 2006a, 116). He is also right to point to the defects of Stephen Jay Gould's "non-overlapping magisteria" compromise between science and religion, as originally formulated (Gould, 2002), and to prefer the formulations of that other agnostic defender of evolutionary theism, Michael Ruse. Attfield's chapter on "Darwinism, Disvalues and Design" concludes with a masterly demonstration that predation, and even parasitism, the phenomenon that so disturbed Darwin himself, "form no significant problem for theists" (Attfield, 2006a, 130), a point to which I shall have to return.

In his "God and Evil" chapter Attfield reformulates the classic problem of evil into the proposition that

> No other world that God could have created would have had a better balance of good over evil than the actual world, despite the many evils it contains, has or will have. (Attfield, 2006a, 135).

This is a very interesting move, because of course it raises the spectre of Dr Pangloss in Voltaire's *Candide*, and Leibniz' best of all possible worlds. Attfield is well aware of the various critiques of best-possible-world theory. These are summarised for instance in Robert Merrihew

Adams' introduction to his edited *The Problem of Evil* (Adams, R.M., 1990). Particularly relevant to my own reservations is Philippa Foot's point, as cited by Adams, that worlds are not good or evil in themselves, but good or evil for particular persons or projects (Adams, R.M., 1990, 7), and D.Z. Phillips' general rejection of the consequentialism implicit in best-possible-world theory as appropriate to talk about God (Phillips, 2004, 35-46). I note also Marilyn McCord Adams' critique that the God of Jews and Christians is not depicted by those traditions as an abstract value-maximiser, but as one who by grace does more for human beings than they are worthy to receive (cf. Adams, M.M., 1990, 210). I am not yet clear that Attfield's formulation above escapes all these problems.

Attfield goes on to show how the Free-Will Defence to moral evil can function within such a proposition. He challenges the famous anti-theodicy of Ivan Karamazov, who would "return his ticket" if all the goods of the world depended on the torture of a single child, as "a disproportionate judgement" (Attfield, 2006a, 137). Attfield then shows convincingly how the Free-Will and Irenaean defences to moral evil can be combined. When he turns to the more difficult problem of natural evil, he concludes that, without evidence, it is hard for us to suppose that a better natural system could exist (Attfield, 2006a, 141). Though widespread disvalue is conceded, its outworkings are seen to be "systemic preconditions of the flourishing of billions of creatures across the ages, as well as of human capacities and of the human endowment" (Attfield, 2006a, 143). There was in Ruse's phrase "no other way" to realise the range of creaturely values we observe other than a world of natural selection, "complete with predation, parasitism, agony and suffering and (apparent) waste" (Attfield, 2006a, 146, drawing on Ruse, 2003, 333).

In my own monograph I pick up this "only way argument" (Southgate, 2008b, 47-48). Attfield and I would both be profoundly wary of the bolder claim that this is the best possible world. I want only to share the more modest claim with which Attfield ends, that a created world realising the sorts of values we observe would have to be a Darwinian world.

It is a thousand pities that Michael Murray wrote *Nature Red in Tooth and Claw* without Attfield's chapters to draw on. As far as I can see Murray never quite grasps the character of Darwinism, and hence he doesn't grasp the force of the "only way" argument as advanced by Attfield, Ruse and myself, among others. Nor does Murray ever quite free himself from variants of anthropocentrism based on the view that animal suffering is at least in part outweighed by the goods arising for human beings. But he does come somewhere near to Attfield's position in his eventual theodicy, which is based on affirming both nomic regularity—the importance to all creatures of living in a consistent world, and a world moreover in which creatures evolve pain responses to noxious stimuli—and also what Murray somewhat oddly calls a chaos-to-order universe (Murray, 2008, 166-92). I suppose that roughly corresponds to a Darwinian universe which, although evolution is not of itself directed, has nevertheless manifested a massive increase in complexity over the last three billion years.

Reservations as to Attfield's Position

So there are points of contact between Murray's analysis and Attfield's, though I think Attfield's is a much clearer and more biologically-informed account. There are however two instances where I think Murray's account picks up something important. First, he cites the issue raised by Marilyn McCord Adams that

> If the good in question requires the evil of treating someone as a means, then no-one has the right to permit that person to be treated in this way, regardless of the supposed greater good it will bring about. (Murray, 2008, 187)

Murray properly points to exceptions within the human sphere, such as quarantining (a sadly topical subject in 2009 with the advent of so-called "swine flu"). But where harms are *systemically* used to promote goods not in those creatures but in others, then there is a charge against the goodness of the exploiter of those individuals, and that is the situation with the harms occasioned by the evolutionary process. This is, in fact, the objection of Ivan Karamazov all over again—if that is how the *system* works, at the expense of the individual sufferer, well then, the protester respectfully and with reason returns his ticket. My own position is that this argument holds in the sphere of evolutionary evil just as in that of moral evil.

Attfield is right to hold that Ivan's argument does not imply that it would have been better for God not to have created at all. Rather I suggest that God's care and love must be operative at the level of every individual, not just in terms of a system that is, on balance, the best that can be achieved. And that leads me on to the second of Murray's insights, which is that evolutionary theodicy must be done using a combination of arguments (Murray, 2008, 193-99). The only way argument in isolation will not do by itself. It fails at the level of the suffering of the individual creature. I would hold that it needs to be embedded within a richer account of the Christian narrative—one that shows that God's care and love is always operative, and makes a difference to the individual creature. In my own formulation this includes the invocation of God's co-suffering with every creature that suffers, and of Christ's atoning work at the Cross, and it also means postulating a fulfilled life for the victims of evolution in some eschatological state.

This I see as my principal difference from Attfield's position on evolutionary theodicy. In Attfield's book he merely says that the possibility of life after death lies beyond the scope of the present work (Attfield, 2006a, 150), so for all I know he will invoke it in some subsequent study—during what I hope will be a long and fruitful retirement. But I am convinced that merely to argue a theodicy on the basis of God as the consequentialist calculator of values against disvalues does as D.Z. Phillips has so trenchantly argued lead to an arid and even possibly an unholy theology (Phillips, 2004). Hence my insistence on offering what Thomas F. Tracy would call a "thick defence" of the justice of God, one which draws on the whole arc of the Christian narrative of creation and salvation, rather than trusting to a single argument (Tracy, 2007, 157-60). The "thin defence" of offering a single logical argument for God's righteousness in the face of evolutionary suffering may succeed philosophically in showing the logical compossibility of God's goodness and evolutionary evil, and I suspect that is Attfield's objective. But I continue to consider that it is vulnerable to a version of the objection of Ivan Karamazov, and what is more that it does little of itself to enrich our understanding of the God of evolutionary creation, as I have tried to do in my more extended treatment (Southgate, 2008b, Chs. 3-5; Southgate, 2011).

The final component of my theodicy is a sense of the high calling of humanity as having a part in the redemption of the non-human creation. It is particularly interesting to me to see how—in both Attfield's work and my own—the way theodicy is framed has implications for ethics. It seems to me that because Attfield is satisfied with this as the world with the optimal balance of value over against disvalue, without the need to invoke other elements such as eschatological fulfilment, so it is natural for him to work on the basis of a stewardship ethic focussed very much on the preservation of the systems that embody this optimal balance. Whereas the implication of my own approach is that this profoundly ambiguous world stands much in need of healing by God, and that after the Cross and Resurrection of Christ we live in the era of that transformative healing. I therefore argue for a co-redeemerly environmental ethic (Southgate, 2008b, Chs. 6-7; see also Southgate, 2009). The remainder of this chapter will be an exploration of some of these distinctions, and I shall then end on some points of agreement.

Issues around Stewardship

It is familiar ground in recent Christian theology that stewardship, as an image for the human vocation in respect of the non-human creation, is both commonplace—almost now the default position—and yet under heavy attack from a range of scholars. For a recent survey see Berry 2006. Particularly important critiques include Clare Palmer's denunciation of stewardship as being (in the sense in which it is popularly used) unbiblical (Palmer, 1992). Anne Primavesi has condemned the concept of stewardship as exploitative and unecological (Primavesi, 1991, 106-7). Sean McDonagh is concerned that "within the context of this analogy the earth is reified and becomes either inert property to be cared for or financial resources to be managed in a way that gives a good return on the investment". (McDonagh, 1994, 130). Edward Echlin claims that stewardship "easily lends itself to a detached and manipulative view of creation" and that it "has not moved hearts" (Echlin, 2004, 16). Bill McKibben regards it as "so lacking in content as to give us very little guidance about how to behave in any given situation" (McKibben, 1994, 51). Beyond that stewardship carries the implicit presumption that there is some state or character of the non-human creation, knowable by humans, that we are in a position to steward. And Attfield does note this criticism, as deployed by Richard Evanoff (Attfield, 2006a, 198-99). *Do* we yet know enough about the Earth to consider ourselves its stewards? Here Attfield helpfully responds by pointing out that a stewardly approach must be combined with vigorous use of the precautionary principle "where irreversible harms or the crossing of ecological thresholds are in question" (Attfield, 2006a, 199).

Attfield addresses some of the other classic criticisms of stewardship in his own article in the Berry book (Attfield, 2006b), in turn an amended version of a chapter in *The Ethics of the Global Environment* (Attfield, 1999). The reasonableness of the tone of this rebuttal is classic Attfield. My reaction on reading this piece is merely to note that the rhetorical flavour of an image such as that of the steward does not equate to its currency in philosophical analysis. The image may be carefully deployed by the reasonable philosopher, and yet continue to carry the wrong connotations when bandied around in the literature and politics of environmentalism. However, Attfield does in my view establish that the image of "steward" (I suspect he rather prefers that of "trustee") is at least consistent with sound environmental practice. While he does not convince me that stewardship "has been a central approach" throughout the Christian centuries (Attfield, 2006b, 84), nevertheless he establishes a plausible continuity between the image of "tending the garden" in Gen. 2 (an image which has found favour with evangelicals, see for example Granberg-Michaelson, 1987), environmental practice in mediaeval Benedictinism, and the explicit adoption of stewardship in the work of Matthew Hale and those who followed him. (Attfield, 2006b, 81-84) His case is that the image need not lead to a sense that the landlord is absent, or that the land is merely a resource to be managed, or indeed exploited or manipulated.

I think it is at least one small pity that my own article on stewardship appeared just at the time Attfield's *Creation, Evolution and Meaning* was itself being published. I would have hoped to persuade him that while stewardship can be one element in a Christian environmental ethic, appropriately deployed in a particular range of circumstances, other emphases are also important (cf. Southgate, 2006). In particular I am concerned that the usual connotations of the word "stewardship" are in terms of caution about the future—*"stewardship" of "resources" seeks to provide a future no worse than the present.* There is an implication here (which most of the relevant ecotheologians would probably hotly deny) that indeed things tend to get worse, and stewardship is important to protect God's Earth because we do not expect its miraculous deliverance or transformation, at any rate any time soon.

Examples of this approach—stewardship as preservation—can be found in Lawrence Osborn's work (Osborn, 1993), also in *Earthkeeping in the Nineties: Stewardship of Creation* by Peter de Vos et al. De Vos et al talk of the calling of the shepherd to "maintain the flock" (de Vos et al., 1991, 292), and note that the

> 'commons' face despoiling, if they are not already spoiled. Perhaps, therefore, the call to stewardship of the 'commons' translates to the establishing of governing bodies, capable of restricting the use of the commons (de Vos et al., 1991, 323).

There is no future hope in this ethic—other than, in some advocates of stewardship, a dubiously biblical hope that one day humans might be able to recreate Eden. Given the sometimes casual talk about Eden in some of the ecotheological literature it is very important to question whether the evidence of the biblical witness does in fact point to a restoration of an initial harmony. Hans Urs von Balthasar for one is clear that it does not. He writes: "The New Testament nowhere speaks of the recovery of a lost glory of the original state, but rather of the eschatological achieving of the righteousness and glory of God in his cosmos" (von Balthasar, 1989, 297; cf. also O'Donovan, 1994, 55-56).

Now Attfield seems to me to invoke in *Creation Evolution and Meaning* a much more active vision of stewardship than some, speaking not just of "preservation" but of "rehabilitation" (Attfield, 2006a, 194) but he still uses the language of the steward as "trustee", as "guardian" (Attfield, 2006a, 193). Not only does this language play into the frequently-levelled charge that it makes God appear to be an absentee owner—trustees or guardians are needed because the normal authority of the owner is absent—but it stresses to a greater extent than I think is appropriate the notion that we are to hold on grimly to what God has left behind.

In some recent work, in collaboration with the New Testament scholars David Horrell and Cherryl Hunt, I have been exploring the implicit cosmological narratives lying behind some ecotheologies, and considering them in terms of Northrop Frye's analysis of narrative genre (Horrell et al., 2010). It seems to me that some of the versions of stewardship to which I have referred are based on an implicit narrative that things always tend to get worse, what Frye would term an ironic narrative. Such a narrative receives biblical support, especially in Qoheleth, and God's answer to Job in Job 38-41—an answer described by McKibben as "deeply sarcastic" (McKibben, 1994, 35), as it mocks the notion that humans can comprehend the ways of the world. An ironic, pragmatic version of stewardship would draw sustenance from that type of vision. But what I argue is that that genre of narrative is hard to reconcile with the implicit cosmologies we find in the New Testament, and with the strong conviction in the Pauline corpus that Christian ethics must be eschatological. If we look at the classic passage in which Paul addresses the non-human creation, Rom. 8. 19-22, I think we can see not only the conviction that creation's current condition is regarded as part of God's hope for the future, pregnant with possibilities, but also that creation's future is tied up with the children of God coming into their freedom, a freedom that is glory.

So this is a narrative of hope, and of the believers' struggle to live out their freedom in Christ, a freedom that in turn participates in the liberating of creation. I strongly question whether stewardship, even what Osborn would describe as strong stewardship (Osborn, 1993), can do justice to this vision. So I have suggested—very controversially—that humans have a part, with God, in the healing of the world. So I side with Ronald Cole-Turner in his suggestion that humans can be co-redeemers with God—as also with his comment that we are at the same time "creatures who constantly stand in need of redemption" (Cole-Turner, 1993, 102). My specific suggestion, again controversial, is that humans in their redeemed freedom might be called to reduce the rate of biological extinction. Thomas Berry has written:

> Extinction is a difficult concept to grasp. It is an eternal concept. It's not at all like the killing of individual life forms that can be renewed through normal processes of reproduction. Nor is it simply diminishing numbers. Nor is it damage that can somehow be remedied or for which some substitute can be found. Nor is it something that only affects our own generation. Nor is it something that could be remedied by some supernatural power. It is, rather, an absolute and final act for which there is no remedy on earth as in heaven. (Berry, 1998, 9; cf. also McDonagh, 2004)

Although extinction has been part of the driver of evolution over the last three and more billion years, it is always as Berry implies a loss to the creation of a whole strategy of being alive, a whole way in which God is praised in God's creation. In this eschatological era a sustained initiative to try and limit any further extinction might be our part in God's eventual healing of the world, the reconciliation of all things in Christ (Col. 1.20).

However, the tragedy of our current predicament is that so far from reducing the rate of non-anthropogenic extinction, we are in the process of engendering, through our elevation of the planet's natural greenhouse effect, a sixth great extinction event, estimated even by a document as cautious as the Stern Report as likely to lead a level of extinction possibly as high as 60% of all mammals (Yohe, 2007, 106). So before we can even begin to think of addressing the levels of non-anthropogenic extinction, there is a vital imperative to address a deepening crisis engendered or at least greatly exacerbated by human activity, that of climate change. And here is where I come back more onto Attfield's ground, because this is an example I believe of where stewardship language can be helpful. Things are indeed getting worse very quickly, and active management to preserve elements of the biosphere, possibly even including major projects in the translocation of species, will surely be necessary. I have recently proposed as a thought-experiment the possibility of moving polar bears to the coast of Antarctica (Southgate, 2009, 262-65).

I was particularly fascinated, when I was reading out Attfield's paper in Crete, to see that he and I were working along similar lines in environmental ethics. Both of us were considering as a priority the need to give voice to the voiceless, to empower those whom the current crisis is rapidly disempowering. Attfield's emphasis in that paper (now published in *The Journal of Global Ethics* (Attfield, 2009b)) is on empowering the poor so that they can also take up the call to stewardship and trusteeship.

And he also has a strong sense of future generations as partners in the project of stewardship—he describes "the true subject of stewardship" as "an intergenerational collective" (Attfield, 2006a, 200). I also have become very interested in the issue of future generations. I am concerned about future human generations as disempowered and dislocated, in some cases losing even the possibility of remaining in their ancient home—one has only think of somewhere like the Andaman Islands, a locus rich in unusual culture and language but much of it likely actually to disappear as ocean levels rise. But I also consider in recent work future generations of non-human creatures which may lose even the opportunity to exist, let alone flourish, as a result of climate change. These, it may be argued, are the new *anawim*, the new poor to whom we should be paying attention as God's special care and concern (Southgate, 2009, 258-60).

I have been arguing strongly and explicitly from the standpoint of Christian ethics, and I have shown that my approach leads me sometimes to a rather different emphasis from Attfield's, but sometimes into some interesting convergences. I would like to note lastly that Attfield's appeal to "secular stewardship" (Attfield, 2006a, 201-2) provides a valuable meeting ground with ethicists of other faiths and none, and may therefore be much more influential in the current context than an appeal to an eschatological Christian ethic. And I think Attfield makes a fascinating observation when he talks of how widespread the impulse to thankfulness is even among secular environmentalists. This, like the impulse to wonder found even in secularists as rabid as Richard Dawkins, is an intriguing hint of how a response to God may lurk even in the most unexpected quarters.

I end by restating how profoundly grateful I am both to Attfield for his work and for the privilege of speaking and writing in his honour.

References

Adams, Marilyn McCord, 1990, 'Horrendous Evils and the Goodness of God', in *The Problem of Evil*, ed. Robert Merrihew Adams and Marilyn McCord Adams (Oxford: Oxford University Press), 209-21.

Adams, Robert Merrihew, 1990, 'Introduction', in *The Problem of* Evil, ed. Robert Merrihew Adams and Marilyn McCord Adams (Oxford: Oxford University Press), 1-24.

Attfield, Robin, 1999, *The Ethics of the Global Environment* (Edinburgh: Edinburgh University Press and West Lafayette, IN: Purdue University Press).

—. 2006a, *Creation, Evolution and Meaning* (Aldershot: Ashgate).

—. 2006b, 'Environmental Sensitivity and Critiques of Stewardship', in

Environmental Stewardship: Critical Perspectives—Past and Present, ed. R.J. Berry (London and New York: T&T Clark Continuum), 76-91.

—. 2009a, Review of *Nature Red in Tooth and Claw*, by Michael J. Murray, Review in *Reviews in Science and Religion*, 53, 36-39.

—. 2009b, 'Ecological Issues of Justice', in *Journal of Global Ethics*, 5 (2) 147-54.

Berry, R.J. (ed.), 2006, *Environmental Stewardship: Critical Perspectives—Past and Present* (London and New York: T&T Clark Continuum).

Berry, Thomas, 1988, *The Dream of the Earth* (San Francisco: Sierra Club Books, 1988).

Cole-Turner, Ronald, 1993, *New Genesis: Theology and the Genetic Revolution* (Louisville, Ky.: Westminster John Knox Press).

Conway Morris, Simon, 2003, *Life's Solution: Inevitable Humans in a Lonely Universe* (Cambridge: Cambridge University Press).

De Vos, Peter, de Witt, Calvin, Dykema, Eugene, Ehlers, Vernon, and Wilkinson, Loren, ed. Loren Wilkinson, 1991, *Earthkeeping in the Nineties: Stewardship of Creation* (Grand Rapids, Mi.: Wm. B. Eerdmans, revised edn.).

Echlin, Edward P., 2004, *The Cosmic Circle: Jesus and Ecology* (Blackrock, Co. Dublin: Columba Press).

Gould, Stephen Jay, 2002, *Rocks of Ages: Science and Religion in the Fullness of Life* (London: Vintage).

Granberg-Michaelson, Wesley (ed.), 1987, *Tending the Garden: Essays on the Gospel and the Earth* (Grand Rapids, Mi.: Wm. B. Eerdmans).

Horrell, David G., Hunt, Cherryl, and Southgate, Christopher, 2010, *Greening Paul: Rereading the Apostle in a time of Ecological Crisis* (Waco, Tx.: Baylor University Press).

McDonagh, Sean, 1994, *Passion for the Earth: The Christian Vocation to Promote Justice, Peace and the Integrity of Creation* (London: Geoffrey Chapman).

—. 2004, *The Death of Life: the Horror of Extinction* (Blackrock, Co. Dublin: Columba Press).

McKibben, Bill, 1994, *The Comforting Whirlwind: God, Job and the Scale of Creation* (Grand Rapids, Mi.: Wm B. Eerdmans Pub. Co.).

Murray, Michael J., 2008, *Nature Red in Tooth and Claw: Theism and the Problem of Animal Suffering* (Oxford: Oxford University Press).

O'Donovan, Oliver, 1994, *Resurrection and Moral Order: An Outline for Evangelical Ethics* (Leicester: Apollos and Grand Rapids, Mi: Wm B. Eerdmans, revised edn.)

Osborn, Lawrence, 1993, *Guardians of Creation: Nature in Theology and the Christian Life* (Leicester: Apollos).

Palmer, Clare, 1992, 'Stewardship: A Case-study in Environmental Ethics', in *The Earth Beneath: A Critical Guide to Green Theology*, ed. Ian Ball et al. (London: SPCK), 67-86.

Peacocke, Arthur, 1990, *Theology for a Scientific Age: Being and Becoming—Natural and Divine* (Oxford: Basil Blackwell).

Phillips, D.Z., 2004, *The Problem of Evil and the Problem of God* (London: SCM Press).

Primavesi, Anne, 1991, *From Apocalypse to Genesis: Ecology, Feminism and Christianity* (Tunbridge Wells: Burns and Oates).

Rolston, Holmes, III., 1999, *Genes, Genesis and God: Values and their Origins in Natural and Human History* (Cambridge: Cambridge University Press).

Ruse, Michael, 2003, *Darwin and Design: Does Evolution have a Purpose*? (Cambridge, Ma. and London: Harvard University Press).

Russell, Robert J., 1998, 'Special Providence and Genetic Mutation: A New Defense of Theistic Evolution', in *Evolutionary and Molecular Biology: Scientific Perspectives on Divine Action*, ed. Robert J. Russell, William R. Stoeger S.J. and Francisco J. Ayala (Vatican State: Vatican Observatory Pubs and Berkeley, Ca.: Center for Theology and the Natural Sciences), 191-223.

Southgate, Christopher, 2006, 'Stewardship and its competitors: a spectrum of relationships between humans and the non-human creation', in *Environmental Stewardship: Critical Perspectives—Past and Present*, ed. R.J. Berry (London and New York: T&T Clark/Continuum), 185-95.

—. 2008a, Review of *Creation, Evolution and Meaning*, by Robin Attfield, *Reviews in Science and Religion*, 51, 33-35.

—. 2008b, *The Groaning of Creation: God, Evolution and the Problem of Evil* (Louisville, Ky.: Westminster John Knox Press).

. 2009, 'The New Days of Noah: Assisted Migration as an Ethical Imperative in an Era of Climate Change', in *Creaturely Theology: On God, Humans and Other Animals*, ed. Celia Deane-Drummond and David Clough (London: SCM Press), 249-65.

—. 2011, 'Re-reading Genesis, John and Job: A Christian's Response to Darwinism', in *Zygon: Journal of Religion and Science*. Forthcoming.

Tracy, Thomas F., 2007, 'The Lawfulness of Nature and the Problem of Evil', in *Physics and Cosmology: Scientific Perspectives on Natural Evil*, eds. Nancey Murphy, Robert J. Russell and William R. Stoeger, S.J. (Vatican City: Vatican Observatory and Berkeley, Ca.: Center for Theology and the Natural Sciences), 153-78.

Von Balthasar, Hans Urs., 1989, *The Glory of the Lord: A Theological*

Aesthetics, Volume VII: Theology: The New Covenant, trans. Brian McNeil C.R.V., ed. John Riches (San Francisco: Ignatius Press).
Ward, Keith, 1996, *God, Chance and Necessity* (Oxford: Oneworld).
Yohe, Gary, 2007, 'Climate Change', in *Solutions for the World's Biggest Problems: Costs and Benefits*, ed. Bjorn Lomborg (Cambridge: Cambridge University Press), 103-24.

CHAPTER SEVEN

REPLY TO CHRISTOPHER SOUTHGATE

ROBIN ATTFIELD

Setting the Scene

Christopher Southgate, whose help and collaboration I have appreciated in recent years, has queried my view that theism and Darwinism can be reconciled without resort to revelation and eschatology, suggesting that a compound theodicy (with more than one strand of defence) is needed. He also criticises what I have written about stewardship. Those who have not been able to read *Creation, Evolution and Meaning* may find these ideas puzzling, and so perhaps I should explain what is argued in the relevant part of that book.

In chapter Six and Seven I argue for the compatibility of Darwinism and belief in creation, but not from scratch. For the case for belief in creation has already been made out in Chapter Five, and the case for preferring Darwinism both to creationism and to variants such as Intelligent Design is now advanced. (To avoid confusion, I should explain that creationism alleges the special creation of species through supernatural interventions, while belief in creation holds that the material world is dependent on God, however the species may have originated; nor need belief in creation commit its adherents to the belief that certain biological phenomena are too irreducibly complex to be explainable by natural selection, as adherents of Intelligent Design maintain. Believers in creation believe in God as author of creation, but need not subscribe to the kinds of interventionism that both creationists and adherents of Intelligent Design tacitly invoke. All this is explained in Chapter Four and recapitulated in Chapter Six.)

The rest of Chapter Six addresses the problems of reconciling phenomena such as predation and parasitism with belief in creation, and argues that they can be reconciled, despite the apparent problems. Chapter Seven addresses various further aspects of the problem of evil, and argues that a law-governed system like the actual system of nature is consistent with what a creator intent on generating a world of value would create. It does not claim (like Voltaire's character Pangloss) that the actual world is the best of all possible worlds, if only because there is no intrinsic maximum for the goodness of worlds. But it does claim that the world's evils are either due to the choices of agents such as human beings or to the operation of systems indispensable for the generation of valuable creatures, non-sentient and sentient, non-human and human.

Theodicy

Southgate recognises that I do not claim that our world is the best possible (despite including "Pangloss" in his title). This I reject for the reason already given. The other reasons suggested by Southgate strike me as inconclusive. Thus some worlds can be better than others (as people realise when they choose between futures). Consequentialism has no application to God, as opposed to supplying a criterion of rightness to human agents with limited knowledge seeking to do what is right in the constrained circumstances of human life. And, while God is not to be seen as a value-maximiser, there must be some relation between divine grace and the emergence of value (value being what there is reason to be glad about). Nevertheless Southgate and I are at one in rejecting Panglossism.

Nevertheless he is right to hold that I claim that worlds can be compared in point of value, and that the actual world is one that a good God could have created. He also takes me to be asserting a related proposition that I was not asserting, and so I need to introduce its context here.

The context was the compatibility of acceptance of the evils implicit in evolutionary processes and belief in God. But we can readily see that there is no inconsistency here, by adding to the disputed combination any possibly true proposition which, in conjunction with some members of the disputed pair or conjunction, implies the remaining one. And this is actually done in *Creation, Evolution and Meaning*, at p. 135. The possibly true proposition is the one now highlighted by Southgate, namely: "No other world that God could have created would have had a better balance of good over evil than the actual world, despite the evils it contains, has or will have." For this, in conjunction with standard theistic beliefs, entails that evils exist, without apparently being self-contradictory; for example, it avoids claiming that the actual world has a better balance of good over evil than any other, as it allows that other possible worlds could have an equally good balance to that of the actual one. Let me emphasise that I was not asserting the truth of this possibly true proposition, and that my arguments do not collapse if it is rejected (for the world could still be compatible with God's good purposes); rather it simply serves to make the logical point just mentioned.

Nevertheless I did conclude, at p. 143, that the actual system of nature "may well be regarded as having an overall balance of value over disvalue". Also, in the context of the possibility of different laws of nature, I wrote, at p. 141, that "we have no reason to believe that a world with a better balance of good over evil than the actual world is possible". With regard to the possibilities under discussion there, I stand by this claim, but it should have had more qualification, since, as I have written in reply to Alan Holland, we can imagine a possible world just like the actual one but better through the addition of one more violet. But this clearly does not mean that God is less than good for failing to create that world. So I can also stand by what I went on to say, that we have no reason to believe "that the actual world is not a world that a good God would create". It may be that Southgate was misled by this passage into supposing that I was asserting the "No other world" proposition of p. 135.

The move made in that passage does, as Southgate later says he suspects my objective to be, "show the logical compossibility of God's goodness and evolutionary evil", without vindicating God's righteousness (another phrase he uses in that passage), as that task would go beyond consideration of God's role as creator. My overall objective, however, was more than the one about compossibility, as I was also replying to the probabilistic argument against the credibility of a good God creating a world such as ours, as well as to inconsistency arguments. My reply to the probabilistic argument is spread out across chapter 7 (and cannot be fully recapitulated here); it focuses on what Southgate calls "the no other way argument", but is supplemented by the argument from value of chapter 8.

Strictly, my argument does not say that there was no other way for God to create the world except though laws of nature such as that of evolution by natural selection. In the closing section of chapter 7, I recognised that a world somewhat resembling the current one could have been brought about through a sequence of miracles rather than through laws of nature, but added that it would be short-lasting without further miracles, and that the circumstances needed for the development of intelligent choices and of virtuous characters would be absent. So I concluded that such a world does not compare in value with one governed by laws of nature such as ours, and that there was thus "no other way" to a world including the flourishing of the creatures of the actual world (human and nonhuman) except through a system involving natural selection. This is a variant of the reply of

Michael Ruse to Richard Dawkins, and as long as the above qualifications are not forgotten, the label of "the no other way argument" can be accepted for present purposes.

Supplementing Natural Selection

Southgate's view is that this argument is vulnerable to objections and is in need of being supplemented. But before considering all this, I should mention what he says about Holmes Rolston and Keith Ward, since it was in the same final section of chapter 7 that I made a slight adjustment to Ruse's reply to Dawkins to accommodate the view, which I had attributed to them as well as to Arthur Peacocke, that a naturalistic supplement to natural selection may be needed to explain the biological phenomena of the actual world. For while natural selection would still be needed, so would the supplementary factors, whatever form they may turn out to take. Southgate does not dissent from this conclusion, but considers Peacocke alone (of the three figures mentioned) to be appealing to additional naturalistic factors, whereas Rolston and Ward, on his interpretation, introduce divine activity to fill gaps and make good what natural selection cannot otherwise explain.

Both Rolston and Ward certainly hold that natural selection is insufficient as an explanation of the actual biological phenomena. Yet it is not clear that the passages cited by Southgate involve them in "inserting God's influence into gaps in the causal order as described by science". Ward's point seems to be that divine agency is the best explanation of "the progress towards greater consciousness and intentionality that one sees in the actual course of the evolution of life on earth" (Ward, 1996, 78); but this claim is consistent with the generic claim of theists that, but for God's creation of nature and of the laws of nature, life and consciousness would not have evolved as they have. It does not commit him to God having intervened during the early stages of evolution on our planet to direct life in the direction of consciousness.

Rolston ascribes to God the information needed "for the key transitions in evolutionary history" (Rolston, 1999, 359), and considers natural selection unable to supply it. While his choice of words leaves open the interpretation that this information was newly inserted by God as life was evolving, it also leaves open the different interpretation that in creating the overall system of nature God provided for this information to be available when it was going to be needed. Southgate is right to be wary of theologians suggesting that natural processes are insufficient, where natural explanations are in question, to generate the effects that we encounter, and must "either be steered or set aside"; and he may be right about one or both of the two theologians under consideration. Yet neither in Ward's case nor in Rolston's must readers take the view that God's activity is being represented as a secondary cause, alongside and on a level with natural selection, rather than as that of the creator of a system of creatures devised so as to "make themselves" (Peacocke, 2004, 142; Attfield, 2006a, 167).

Objections to the "No Other Way" Argument

Let us return to Southgate's reservations about the "no other way" argument, an argument which he seems to endorse when he agrees that "a created world realising the sorts of values we observe would have to be a Darwinian world". Yet he raises the objection that if attaining goods involves treating someone as a means, that is unacceptable, at least if it is done systematically, and this gives the protester reason to "return his ticket". However, as Ruse has pointed out, a Darwinian world is a world of predation and parasitism, for that is what natural selection involves; and apart from a world of supernatural interventions, the only alternatives open to a creator are a world of unicellular creatures or a world without life at all, or so the "no other way" argument concludes. So people who go along with it (as Southgate does) have no choice but to accept predation and parasitism, rather than to return their ticket. For to grant this argument is to grant that a creation is acceptable in which some creatures suffer as part of a system which facilitates or brings about goods such as animal and human flourishing; for the evolutionary processes of natural selection comprise such a system.

By now, what is left of the original objection is the suggestion that harms of the kind just described amount to God treating someone as a means (to the good of others). But this language is not obviously applicable. Certainly in inter-human contexts there are ethical objections to treating another human being as a means, although, as Michael J. Murray (2008) admits, this may be justified in cases of quarantine. But these problems arise in cases where the human being could have been treated otherwise, and where doing so would have respected their autonomy. Creating species some of which will become extinct through systemic natural forces is another matter, partly because of their lack of such autonomy, and partly because their very existence depends on these same forces, and there is no such thing, as long as natural forces remain in place, as these species being treated otherwise. Thus to suggest that ammonites were treated as a means because they became extinct and that

God should either not have created them or should have intervened to prevent their extinction is out-of-place. Given the options of the actual system of life or a world of unicellular creatures only or one of no life at all, the language of means becomes inappropriate.

But what of individual creatures that suffer in the course of natural selection, for example, frogs eaten by pythons? The suggestion that they (and all the other creatures eaten by the same pythons) are treated as means also seems inappropriate, because if they were not eaten, there could be no pythons, and generally if there were no predation there could be no higher predators, and probably no humanity. So the alternatives, if nature is to be governed by natural regularities, are a world of predation and a world either without life or with unicellular creatures only. But a creator who adopted either of the latter pair of options could with at least as much cogency be accused of treating as means all the creatures that would have existed had predation been authorised, through declining to create them.

Southgate, however, also represents appealing to the "no other way" argument without appealing to specifically Christian claims (such as God's co-suffering with every suffering creature, and generating "a fulfilled life for the victims of evolution in some eschatological state") as making God a "consequentialist calculator of values against disvalues", the charge against such theodicies of D.Z. Phillips (2004). Phillips, it should at once be said, also regarded Christian eschatology involving life after death with disdain, but the issue still arises of whether adherents of the "no other way" theodicy should accept his "consequentialist calculator" charge.

Here it should be replied that God would have sufficient knowledge not to need to calculate, but could simply judge which possible world was worth creating; and in doing so would presumably take into account values and disvalues, rather than ignoring them. (If not, the critic comes close to saying that there are no possible worlds, however strewn with tragedies and atrocities, which God would decline to create.) The theory of consequentialism (as was mentioned above) is needed for us to understand the rightness of human actions, rather than God's, and applies where relevant consequences are restricted to (humanly) foreseeable ones rather than ones that God alone could know. (Nor does the consequentialism that I uphold require even human agents to calculate the consequences of each and every action; in its more plausible versions, it works from the causal tendencies of types of action and/or of motivation, and declares behaviour right which complies with the types with the best overall impacts in general, as is argued in my response to Clare Palmer.) Accordingly the charge of making God a "consequentialist calculator" is no more in place than the charge that God treats suffering creatures as means, and should not be allowed to count against the "no other way" argument, whether supplemented or unsupplemented. (Phillips' criticisms of other theodicists cannot be considered here.)

In the same connection, Southgate suggests that "God's love and care must be operative at the level of every individual". Here it may be helpful to ask which claim it is for the sake of which this "must" applies. For it would be one thing for this to be required if New Testament claims about God are to be reconciled with Darwinism and the evils that Darwinian processes involve, and quite another where what is in question is the compatibility of Darwinism and the kind of theism that is common to Christianity, Judaism and Islam (or undifferentiated theism). What Southgate says may be necessary for the reconciliation of the former pair, but is less cogent as a requirement of the latter combination. For God could be good and loving in conferring on all actual creatures existence in a world of great intrinsic value, without conferring on them guaranteed security or long-term immunity from the processes by which that world is run. (Indeed many Christians would assent to this view of God's goodness, without insisting on a stronger sense involving such immunity.)

Southgate, however, in presenting an eschatology in which non-human animals are included (see also Southgate, 2008), implicitly appeals to revelation, and is unimpressed with attempts to reconcile Darwinism and belief in God that stop short of this. Here I would like first to clarify the stance of *Creation, Evolution and Meaning*, which makes no appeal to revelation, and declares such matters beyond its scope. This does not mean that I have no time for revelation, but that I was attempting to see to what extent these beliefs and their co-tenability can be based on reasoning accessible to believers and non-believers alike. Thus I was not attempting to reconcile with Darwinism the claims of the New Testament, but rather the theism that is common to Judaism, Christianity and Islam. So my project is rather different from Southgate's, and its implicit methodology is correspondingly different.

On the subject of revelation it should be remarked that I have also argued, in my review of Murray's book, that appeals to revelation cannot be used to give grounds for belief in God's goodness, for belief in God's goodness is a presupposition of there being such a thing as revelation at all (Attfield, 2009). What is revealed could still, of course, supply additional grounds.

I should add that the difficulty with the move of invoking an animal heaven is to find grounds for this belief, and for that reason I have preferred to stick to beliefs not dependent on revelation, even where ancillary and possibly true beliefs would make the task easier. Thus partly for the same reason I avoid appealing to the belief, held by some, that evils such as premature death resulting from cancer are to be ascribed to the free choices of Satan or Satan's cohorts, who have somehow been put in charge of the relevant natural regularities (a belief seriously considered by Murray). This ancillary belief would not, in my view, help to make a theodicy more credible, but rather less credible, as it would suggest massive bad planning on the part of the creator in giving this authority to such free spirits. I am not suggesting that Southgate's additional beliefs share this liability, but I prefer to argue from premises available to all-comers, as is appropriate in a work intended for a secular readership as much as for a readership of believers.

Such reasoning, I claim, supplies a clear basis for resisting both the view that theism and Darwinism are probably incompatible or not readily co-tenable, and the view that they are probably incompatible until and unless we are given some further premises which tell us that the suffering of humans and of nonhumans alike will be compensated in the life to come. At the same time, I want to resist Southgate's view that this theodicy implies the sufficiency of an ethic of preservation rather than one of transformative healing or of social justice. Even if these themes do not emerge from my endorsement of stewardship, they figure prominently in my writings about ethics (such as Attfield, 1995, 1999 and 2003). Besides, these themes and the theodicy that I present (including as it does a version of the Free Will Defence) are not unconnected; for, if this Defence stands up, the free choices facilitated and desired by the creator will include actions through which the needs of fellow-creatures are met, and at the same time virtuous traits are developed in their agents, none of which would have been possible if this autonomy had been absent.

Stewardship and Hope for the Future

On the subject of stewardship, Southgate reports criticisms from others, and adds some of his own. I have replied in some detail to Palmer's critique in *The Ethics of the Global Environment* (1999), chapter 3, a chapter reprinted in R.G. Berry's *Environmental Stewardship* (Attfield, 2006b). For example, I contested the claim that stewardship beliefs represent God as an absentee landlord. Several of the further criticisms that Southgate proceeds to mention represent stewardship as an anthropocentric and manipulative stance, incapable of recognising the value of the natural creatures for which the stewards are to care; but this is an unhistorical interpretation, which ignores a long history of non-anthropocentric adherents of stewardship from the early Modern periods right up to the present (as I argue in *The Ethics of Environmental Concern* (Attfield, 1983 and 1991)). As for Bill McKibben's claim that stewardship beliefs supply too little guidance (McKibben, 1994), my response is that stewardship beliefs are not to be expected to serve as a detailed ethic, rather than as a model providing for human self-understanding; that is why I have gone on to present an environmental ethic, as in works like *Environmental Ethics* (2003).

Southgate now claims that, reasonable as my defence of stewardship in Attfield, 1999 and 2006b may be, stewardship's image continues to have the wrong connotations. So be it, if that is the case, for I am not in the business of image-enhancement, as opposed to exploring implications and presuppositions. He further suggests that stewardship beliefs imply that we are seeking a future no worse than the present, and a background belief that things tend to get worse. To this, I want to reply that stewardship (for example, of effort and resources) need by no means imply that the future cannot be better than the present. The world is unduly full of inhabitants with unsatisfied basic needs, and needs action and policies for development (preferably sustainable development, since unsustainable development will let down the following generations), which need not be incompatible with stewardship. The belief that things tend to get worse would be liable to undermine such effort, but fortunately is no implication of stewardship any more than the myth of the non-improving future is. Once again, what is needed is an ethic relevant to intra-human as well as environmental issues.

Clearly there is much more to Christian ethics than stewardship, although many Christians acknowledge it as one component of Christian ethics; but after all I was not seeking to expound Christianity. However, the view that humans have a part to play in the healing of the world is one that I am happy to endorse; I would merely want to add that it is not incompatible with a model of stewardship, or with a consequentialist ethic either. The further view, though, that through the sufferings of Christ God participates in the world's sufferings and in its redemption, is one that is unavailable to me, and I have to leave it for those who manage to believe in the incarnation of God in Christ, and in the ability of God to suffer.

Southgate has some final remarks on the extinction of species, and on humanity being called to reduce its rate. Here, I agree that it is sometimes the role of humanity to preserve species that would otherwise become extinct, as has been done with the Arabian oryx. I am also interested (indeed intrigued) at his suggestion that (some) polar bears should be moved to the Antarctic. However, I am reluctant to agree that we should do all that we can to prevent extinctions. For extinctions are inherent in natural selection, and the very natural processes that have allowed humanity to emerge have also involved the loss of a large proportion of all the species that there have ever been. Adopting an evolutionary theodicy surely involves accepting the inevitability of at least some extinctions. Even if it were possible to resuscitate the mammoths and the dinosaurs, that surely would be low on the list of our current priorities.

These priorities include meeting human needs, preserving peace, equitably stabilising the climate, and bequeathing a sustainable world to our successors. These are far from hopeless undertakings. Indeed hope is a presupposition of consequentialist ethics and of Christian theology alike, as Southgate and I could readily agree.

References

Attfield, Robin, 1983, *The Ethics of Environmental Concern* (Oxford: Blackwell).

—. 1991, *The Ethics of Environmental Concern*, 2nd edn. (Athens, GA and London: University of Georgia Press).

—. 1995, *Value, Obligation and Meta-Ethics* (Amsterdam and Atlanta: Éditions Rodopi).

—. 1999, *The Ethics of the Global Environment* (Edinburgh: Edinburgh University Press, and West Lafayette, IN: Purdue University Press).

—. 2003, *Environmental Ethics: An Overview for the Twenty-First Century* (Cambridge: Polity Press and Malden, MA: Blackwell).

—. 2006a, *Creation, Evolution and Meaning* (Aldershot and Burlington, VT: Ashgate).

—. 2006b, 'Environmental Sensitivity and Critiques of Stewardship', in R.J. Berry (ed.), *Environmental Stewardship: Critical Perspectives—Past and Present* (London and New York: T&T Clark/Continuum), 76-91.

—. 2009, Review of Michael J. Murray, *Nature Red in Tooth and Claw*, *Reviews in Science and Religion*, 53, 36-39.

McKibben, Bill, 1994, *The Comforting Whirlwind: God, Job and the Scale of Creation* (Grand Rapids, MI: William B. Eerdmans).

Murray, Michael J., 2008, *Nature Red in Tooth and Claw: Theism and the Problem of Animal Suffering* (Oxford: Oxford University Press).

Peacocke, Arthur, 2004, 'Articulating God's Presence in and to the World Unveiled by the Sciences', in Philip Clayton and Arthur Peacocke (eds), *In Whom We Live and Move and Have Our Being* (Grand Rapids, MI and Cambridge, UK: William B. Eerdmans), 137-154.

Phillips, D.Z., 2004, *The Problem of Evil and the Problem of God* (London: SCM Press).

Rolston, Holmes, III, 1999, *Genes, Genesis and God: Values and Their Origins in Natural and Human History* (Cambridge: Cambridge University Press).

Southgate, Christopher, 2008, *The Groaning of Creation: God, Evolution and the Problem of Evil* (Louisville, KY: Westminster John Knox Press).

Ward, Keith, 1996, *God, Chance and Necessity* (Oxford: Oneworld).

CHAPTER EIGHT

PERFECTIONISM AND HEDONISM

ROGER CRISP

I have admired and learned from Robin Attfield and his work since my days as a graduate student in the 1980s. Philosophical insight, quickness of thought, open-mindedness, and generosity are paradigmatic virtues of a professional philosopher, and Attfield possesses these, and others, in abundance. He was kind enough to examine my doctoral thesis on ideal utilitarianism in Oxford in 1988, along with Sir Geoffrey Warnock. I shall never forget the look of wonder on Sir Geoffrey's face as Attfield and I, in discussing one of Derek Parfit's thought experiments about future generations, earnestly debated whether chimpanzees could appreciate muzak. It is a pleasure and a privilege to be asked to contribute to this volume in his honour.

Attfield has developed a world view in which his ethics is situated, and I am sympathetic to several components of both, including his broadly consequentialist outlook and his cognitivism. I am more inclined towards non-naturalism than he is, since naturalism—understood as the view that the world should be understood to serve as the subject only of those properties essentially predicated of it in natural scientific explanations—seems to me to strip the world of the kind of normativity underlying the kind of cognitivism and consequentialism Attfield finds plausible. But I suspect that our differences here may be only apparent, and that Attfield may be working with a broader account of naturalism that would allow for the kinds of evaluative and normative property I prefer to characterize as non-natural. But there is one issue on which we do clearly disagree, and on which my views have over the years moved further from Attfield's: the good, and in particular the "good for", understood as equivalent to "flourishing" or "well-being" and as what underlies the moral standing or considerability of any being in possession of it.

According to Attfield, the common restriction of moral standing to sentient creatures in the western tradition is unjustified (Attfield 1995, 20-

21; all unattributed references in the text are to this book). This is because non-sentient beings have a good of their own and can "flourish after [their] own kind". The reference to "kind" here brings out Attfield's *perfectionism*, according to which the good of any being consists in its "perfecting" those essential qualities that make it the kind of being it is:

> One of the elements present in the flourishing of members of a species ... consists in the development of those potentials in the absence of which from most of its members a species would not be recognizable as the species it actually is in our world, rather as creatures otherwise resembling pigs but equipped to fly rather than to walk and trot would neither be recognizable as pigs nor *be* pigs. (48)

As Attfield notes, perfectionism underlies Aristotle's position, and the question of how to define a being's *ergon* ("function" or "characteristic activity") was as difficult for him as those of what a species is or how to draw the boundaries of species are for us. Now the pig is in fact not usually seen as a species at all, but a genus. Imagine that the flying animals Attfield imagines here could mate with ordinary, terrestrial pigs, to produce flying offspring. I suspect we might well be tempted to conclude that the flying animals were indeed a species of pig. But imagine now that they could not interbreed, and that the flying pigs (which we must presume have legs and toes, since they do, according to Attfield, resemble the pigs we know) had an odd number of toes. That would put the flying animals into a completely different order of mammals from cloven-hoofed pigs, sheep, and so on. At this stage of zoological classification, however, it might seem that we have moved some distance from the idea of flourishing (do we really need one account of flourishing for the even-toed ungulates, and another for the odd-toed?), and I shall suggest below that this is indeed the case.

The "potentials" Attfield has in mind in the quotation above are *essential capacities*, viz., those capacities which are essential to the identity of any particular species. These may, of course, include capacities which are not *distinctive* of the species in question, but are shared across others, perhaps many other, species (49).

Also important here, according to Attfield, is that, properly to constitute flourishing, the development of essential capacities should be "harmonious" (53-4). How are we to understand the idea of "harmony" here? On one conception, it is purely aesthetic—a combination of elements which is pleasing to an appropriately sensitive observer. But this seems to me to introduce an *unharmonious* aspect into an otherwise zoologically-based perfectionist theory! On another conception, the harmony is to be understood in terms of the relation of the capacities to one another. So the development of capacities as a whole could be said to be unharmonious to the extent that the development of one capacity hinders the development of another (so in the case of a flying pig, for example, time spent flying could stunt the growth of the muscles required for terrestrial ambulation). But here harmony does not in itself contribute to flourishing. Flourishing consists only in the development of essential capacities, and disharmony is merely an impediment to that.

What, then, does human flourishing consist in? At its most basic level, physical health, including the capacities for growth and self-motion (38-9, 75). At a higher level, the exercise of the senses (55), and mental sanity, including responsibility for one's attitudes and actions, as well as practical rationality, memory, and the emotions (56), meaningful work (57-9), autonomy (63), self-respect (65), aesthetic appreciation (67), and friendship (67-8), and self-creation (68).

Attfield's view of what is "good for" beings is unusual in contemporary ethics, which is dominated by desire theories of well-being and objective list theories which do not include "lower-level" capacities shared with plants (see Parfit 1984, appendix I). Attfield also disagrees, of course, with philosophical hedonists, and I shall come to that view later in this paper. What are Attfield's main arguments for extending moral standing to non-sentient, living beings? He suggests that drawing the line at sentience conflicts with some quite common intuitions, such as those of John Rodman:

> I need only to stand in the middle of a clear-cut forest, a strip-mined hillside, a defoliated jungle, or a dammed canyon to feel uneasy with assumptions that could yield the conclusion that no human action can make any difference to the welfare of anything but sentient animals. (Rodman 1977, 89; quoted at 20)

Most people would also accept the argument—which we might call the "last person argument"—that it would be wrong for the last person on earth to cut down a tree for no good reason (21-2) (see Routley 1973; also Attfield's discussion of the Moorean "two worlds" case, at 22).

There is also some confusion about the notion of flourishing (21). If we allow that plants can flourish, we do not have to accept that cars can too, since talk about what is good for cars can be reduced to talk about what is good for their owners. This is not the case with, say, trees, and indeed their interests can conflict with those of sentient beings, as when they grow so as to obstruct footpaths.

Now, as far as intuitions go, they can equally be marshalled in favour of the sentience criterion. Consider the following from Jonathan Glover:

> If, travelling in a train through the middle of a ten-mile railway tunnel, I saw a man leaning out of the window into the darkness, I might wonder what he was doing. If it turned out to be G.E. Moore spraying the walls of the tunnel with paints, because painted walls are better than unpainted ones, even if no one ever sees them, I should not be able to prove him irrational. But I should not accept his offer of the use of a second paint spray, except possibly out of politeness. (Glover 1984, 110)

How should we proceed in philosophy when fundamental intuitions conflict in this way? According to Henry Sidgwick:

> Since it is implied in the very notion of Truth that it is essentially the same for all minds, the denial by another of a proposition that I have affirmed has a tendency to impair my confidence in its validity ... And it will easily be seen that the absence of ... disagreement must remain an indispensable negative condition of the certainty of our beliefs. For if I find any of my judgments ... in direct conflict with a judgment of some other mind, there must be error somewhere: and if I have no more reason to suspect error in the other mind than in my own, reflective comparison between the two judgments necessarily reduces me temporarily to a state of neutrality. And though the total result in my mind is not exactly suspense of judgment, but an alternation and conflict between positive affirmation by one act of thought and the neutrality that is the result of another, it is obviously something very different from scientific certitude. (Sidgwick 1907, 342)

I can see a strong case for the view that, in the stand-off between Attfield and Glover here, we should suspend judgement. But on reflection I cannot help but find the sentience view more plausible than Attfield's more encompassing position, and I do think that there may be a case for suspecting error in his position. This is because there is an explanation available for the intuitions underlying his view—an explanation which as a consequentialist he should be sympathetic towards, and which throws doubt on those very intuitions. Plants are invaluable for human beings. Without them, we could not exist, since they contribute essential elements to our environment and our health. We gain great enjoyment from eating them and in some cases from contemplating them. So it is easy to see how a group of beings that placed special value on plants might do better than a group which did not and was prepared to sacrifice them for any short-term benefit. An account of cultural evolution, in other words, may have the resources to debunk the intuitions underlying Attfield's perfectionism.

But there are anyway other problems with the position. First, the same sort of problem arises for Attfield's position as for Aristotle's in the equation of well-being with flourishing as a member of a species (see Glassen 1957). We may accept that some plant is an outstanding example of its kind, and is in that sense "good". But the question remains whether being like that is good *for* the plant. There are further intuitions to which we might appeal in support of sentience here. Let me adapt an example from a book of mine on J.S. Mill:

> You are a soul in heaven waiting to be allocated a life on Earth. It is late Friday afternoon, and you watch anxiously as the supply of available lives dwindles. When your turn comes, the angel in charge offers you a choice between two lives, that of an oyster, or that of a plant. The oyster's life will consist only of mild sensual pleasure, rather like that experienced by humans when floating very drunk in a warm bath. When you request the life of the oyster, the angel sighs, 'I'll never get rid of this plant's life. It's been hanging around for ages. Look, I'll offer you a special deal. The oyster's life will last ten years. But I'll make the plant's life as long as you like'. (Adapted from Crisp 1997, 24)

Now I accept that there are some difficulties here in understanding what it would be for *me* to live the life of a plant, or indeed an oyster. But all we need to grasp here is the idea of *what it would be like* to live the life of an oyster (rather pleasant) and that of a plant (nothing at all). Many will feel, when confronted with this example, that there is *nothing* to be said, in terms of well-being, for the life of the plant, however long it is.

There are also concerns about the language of flourishing extending beyond that of individual living members of species. Attfield may be right that talk of what is good for cars can be parsed into language referring to the interests only of their owners, and others. But it is far from clear to me that this can be done with, say, species themselves (24) or capitalism (27). These items flourish in ways quite similar to individual living beings: their lives begin at a certain point; they grow, develop, and change; they face certain threats, which they can overcome; they can die. I suggest that it is only appeal to the welfarist notion of what is "good for" some individual, resting on the conception of what it would be like to be such a being, that will enable us to rule out species and political systems as having moral standing, and this notion will then rule out individual plants as well.

I suggest also that there is a problem in the direction of analysis of perfectionist positions. According to the perfectionist, we should first seek an impartial, "value-free" account of a being's essential and species-specific nature. But there is a serious danger of the perfectionist allowing her conception of well-being to guide her account of nature. Consider, for example, the remarkable number of perfectionist philosophers over the centuries (Aristotle among them) who have claimed that it is part of our nature to philosophize! This charge could be made against Attfield's position. It is as plausible to claim that human beings are characteristically inclined to rest and amusement as it is to claim that it is in their nature to engage in meaningful work; or that philistinism is typical of human beings as that it is in their nature to engage in aesthetic appreciation. Even if these accusations do not stick, we can still imagine beings *just like us* except that they are characteristically lazy and vulgar (though they have the capacity to accomplish great things and admire great works of art and nature). Why should we think their good depends in the slightest on their fulfilling their nature? Further, just as it is one of our capacities to grow, so it is also true that we all have the capacity to age and die. These are as irrelevant to our well-being as are growth, birth, and life, considered in themselves.

Let me now turn to hedonism, which is in a way at the very opposite end of the spectrum of views of flourishing and well-being from Attfield's. Not only will hedonists rule out physical health and other bodily capacities from well-being, but they will include every other component of the good listed by Attfield only in so far as they increase the pleasure or enjoyment, or diminish the pain or suffering, of the individual in question.

Here Attfield appeals to Robert Nozick's example of the "experience machine" (Nozick 1974, 42-5):

> We are offered the opportunity of being wired up to a machine which gives those attached to it a series of exclusively enjoyable experiences, though in fact the experiences would be caused by stimulation of the brain, and the person having them would be passively floating in a tank. Given the chance to go on the Experience Machine for any substantial period of time, most people would refuse, disclosing that they value something other than enjoyable experiences. Their reason for this might lie in the value of autonomy, in the strong sense of the ability to act off your own bat, and for reasons of your own; and thus in the value of the exercise of practical reason, something which would be foreclosed by life on the Machine. Or the reason might lie in the loss of other powers, such as the ability to mould your own life. (38-9)

Partly because of the force of Nozick's example, hedonism suffered a dramatic decline in popularity during the twentieth century. In the remainder of this paper, I shall try to defend it against the experience machine objection, in the hope of persuading Attfield at least that hedonism cannot be dismissed as quickly as it usually is by reference to Nozick (see also Crisp 2006, 117-125).

Let us take as our example that of someone (let's call them P) *actually* writing a great novel, autonomously and through the exercise of practical reason, as compared with someone else (Q) who is wired up to an experience machine and merely has P's experiences "played back" to her. The first point to note is that accomplishing some goal of this kind is often, both in process and in outcome, hugely enjoyable for the person concerned. Indeed, most of the alleged goods cited by non-hedonists do tend to be things we usually enjoy. It would be considerably more worrying for the hedonist if certain items were considered components of well-being though entirely independent of pleasurable experience. Relatedly, we should remember that aspect of human psychology which underlies the so-called "paradox of hedonism". Someone who strongly believes in the non-instrumental value of accomplishment may well be more strongly motivated to pursue such a goal than a person who accepts hedonism. Thus we can see that, over time, cultural evolution may have led human beings to develop evaluative dispositions and understandings of goods which, though in terms of their content they are non-hedonistic, are in fact based on their capacity for promoting pleasure and enjoyment. How might this have happened? One likely source for important components of our value systems is the Stone Age of the hunter gatherers in Europe, Asia, and Africa, which ended as recently as 4000 BCE. Those who achieved more in the field—who brought back bigger bags of food—would almost without doubt have been offered esteem and status within the group. Similar accounts could be told of those who showed autonomous initiative, perhaps.

A different line of argument runs in parallel with the case of the oyster and the plant discussed above. Consider the life of someone—let's call her R—whose life is very similar to that of P, except that all enjoyment (and suffering) has been stripped out. R also writes a great novel, perhaps motivated by a strong sense of aesthetic duty, but gets no enjoyment out of what she is doing, or the accolades she receives for her work. Unlike in the case of the plant, there is certainly here *something that it is like* to be R. But is this a life which is really *good for* her, as opposed perhaps to a life which well illustrates the development of certain important rational capacities often possessed by human beings (capacities the exercise of which is often enjoyed, but need not be)?

There are also concerns about perspective in the case of many non-hedonic values, including accomplishment and the exercise of autonomy. Perfectionist views tend to be somewhat anthropocentric, appealing to our views about what is significant in typical human lives. But, as Thomas Nagel puts it, "In seeing ourselves from outside we find it difficult to take our lives seriously" (Nagel 1986, 214). P might consider her novel "from the point of view of the Universe", perhaps, and see its significance dwindle before her eyes. Consider all the other novels that have been and will be written; consider all the other great works achieved by humanity; then compare all this to the universe itself, with all its beauty and profundity. Or imagine that we could all write as well as Tolstoy (see Nozick 1974, 241, 245). Our benchmark for achievement is arbitrary. Now it could be that these issues of perspective can satisfactorily be resolved. But they are problematic for proponents of non-hedonistic values, and not for hedonists (who make no claims about significance or value "from the point of view of the universe"). Similarly, free will poses serious difficulties for non-hedonists. The ascription of value to autonomy or accomplishment rests on the plausibility of the idea of free will. But both libertarian and compatibilist conceptions of freedom are notoriously problematic. Again, non-hedonists are obliged to provide a defence of free will, while hedonists are not (enjoyment is valuable whether we are free or not).

Robin Attfield's work has advanced our understanding of many of the key questions in moral philosophy. In this paper, I hope to have shown that his own answer to one of these questions—concerning the scope of moral standing and the idea of "good for"—faces certain objections, and that a more traditional hedonist answer to this question has more to be said for it than he and many others in contemporary philosophy are inclined to believe. I can think of no better words to end with than those with which Attfield concludes the 'Introduction' to *Value, Obligation, and Meta-ethics*:

> In the end... such beliefs must each depend on their own grounds, and works of philosophy must stand or fall by their arguments. I am content for the current study to be judged on this basis.

References

Attfield, R., 1987, *A Theory of Value and Obligation* (London, New York, and Sydney: Croom Helm).

—. 1995, *Value, Obligation, and Meta-ethics* (Amsterdam and Atlanta, GA: Rodopi). This book is a development of *A Theory of Value and Obligation*.

Crisp, R., 1997, *Mill on Utilitarianism* (London: Routledge).

—. 2006, *Reasons and the Good* (Oxford: Clarendon Press).

Glassen, P., 1957, 'A Fallacy in Aristotle's Argument about the Good', *Philosophical Quarterly*, Vol. 7, No. 29, 319-22.

Glover, J., 1984, *What Sort of People Should There Be?* (Harmondsworth: Penguin).
Nagel, T., 1986, *The View from Nowhere* (New York: Oxford University Press).
Nozick, R., 1974, *Anarchy, State and Utopia* (Oxford: Blackwell).
Parfit, D., 1984, *Reasons and Persons* (Oxford: Clarendon Press).
Rodman, R., 1977, 'The Liberation of Nature', *Inquiry* 20, 83-145.
Routley, R., 1973, 'Is There a Need for a New, an Environmental, Ethic?', *Proc. of the 15th World Congress of Philosophy*, Varna, Bulgaria: XVth World Congress of Philosophy, 205-10.
Sidgwick, H., 1907, *The Methods of Ethics*, 7th edn. (London: Macmillan).

CHAPTER NINE

REPLY TO ROGER CRISP

ROBIN ATTFIELD

Roger Crisp's acknowledgement of my work can gladly be reciprocated, as I too have greatly benefited from his work. Indeed we may agree more widely than seemed the case, as I am a naturalist not in the sense that he rejects near the start of his paper, but in a sense with which he, as a cognitivist, could well sympathise. (See *Value, Obligation and Meta-Ethics,* 1995 (hereafter VOME), 208-253.)

His paper, however, concerns hedonism and perfectionism, a term used here of theories of the good that relate to the development of capacities (and not in the different sense employed by Derek Parfit: see VOME, 170). As Crisp says, my variety relates to the development of essential capacities (in the sense of "essential" that he presents). But it does not, as he suggests, relate to "perfecting" these essential capacities, as opposed to developing them. My position should not be construed as involving perfectionism in the ordinary sense of the term, about the avoidance of which seminars are these days run by University managements.

Around here, Crisp asks whether we need different accounts of the flourishing of similar creatures that happen to belong to different species, differing in the number of toes they have (his example). Here I would say "yes" if the different number of toes is matched by any difference of essential capacities, such as differences in speed of running or adeptness at climbing or rooting or sure-footedness, for these differences correspond to the availability of different skills or competences, and thus flourishings of different kinds. (More could be said about the relation of essential capacities and flourishing to species boundaries, but this is not the place to do so.)

Crisp problematises my requirement for a creature's flourishing that the development of essential capacities be harmonious (VOME, 53-54). Of his suggested interpretations, I clearly did not intend the first (a combination of elements which is pleasing to an appropriately sensitive observer). His second suggestion is better on target: "unharmonious" capacity-development would here involve the development of one capacity hindering that of another by taking away time that could have been spent on its development; but as he says, harmony in this sense does not add to flourishing independently of the actual development of the various capacities. Yet this would not be the case if unharmonious (or disharmonious) capacity-development involved the development of one capacity taking a form that actually frustrated or prevented the development of another (or of others), as when the development of someone's gustatory capacities produces obesity which prevents the development of athletic ones; by contrast with this, harmonious capacity-development would involve the development of each capacity taking a form consonant with the that of (at least most of) the others. In this sense, harmoniousness would contribute to flourishing. So I continue to think that this criterion of flourishing has a useful role.

Crisp now tackles my case for recognising the moral standing of all living creatures (which ties in with their good turning on the development of their essential capacities), and reports some of the thought-experiments that support this view (but omits others such as that of Donald Scherer (VOME, 22, not to mention the analogical argument mentioned on the same page and on p. 20)). He then quotes an amusing passage from Jonathan Glover, representing it as supporting intuitions conflicting with mine. But there is no conflict here at all; I can readily agree with Glover that painted but unseen railway tunnels are not intrinsically better than unpainted unseen railway tunnels. (My view of aesthetic value is given at VOME, 22.) Almost certainly Glover (with his mind-dependent theory of value: see VOME, 21) would disagree with Routley, Goodpaster, Scherer and me at some stage, as generally would sentientists; but where intuitions are concerned, the examples cited in Crisp's essay are either supportive of my view or, like the one from Glover, neutral.

However, Crisp now supplies an evolutionary argument to explain the intuitions about plants that support my view, suggesting that such "an account of cultural evolution… may have the resources to debunk the intuitions underlying Attfield's perfectionism". But an account like this of how people could have begun recognising instrumental or inherent value in trees does not serve to debunk current beliefs in trees having intrinsic value. (The concept of inherent value is explained in *The Ethics of Environmental Concern*, 1983 and 1991, at pp. 151-2.) One problem is that entirely different kinds of value are concerned. But another is that quite generally evolutionary accounts of tendencies to adopt beliefs fail to debunk those beliefs. A good example is our hypothetical Hypersensitive

Agency Detection Device, which inclines us to interpret ambiguous evidence as involving the presence of an agent, and would probably have been adaptive in saving us from hidden predators and assailants. This device is sometimes adduced as explaining the prevalence of religious belief, and undermining it at the same time. But a range of scholars have recently recognised that the soundness of the HADD hypothesis is fully compatible with the truth of (at least some of) the religious beliefs that some theorists take it to debunk (see Schloss and Murray, 2009); explaining the origins of a belief often fails to explain it away. (Indeed if things were otherwise, it would be open to hedonism's opponents to retaliate by supplying an evolutionary account of intuitions supportive of hedonism, in case that served to undermine people's tendency to uphold it; but actually such a move would be equally inconclusive.)

Crisp now directs attention to problems about whether my account of flourishing works with what is *good for* (say) a plant. My view is that for a plant, growing is far better than languishing from lack of nutrition or being stunted by injury or disease. Here Crisp replies that there is nothing which it is like to be a plant, because of plants' lack of consciousness, and that for this reason no one would choose the life of a plant (as opposed, even, to the life of an oyster). But these facts have no tendency to show that plants lack moral considerability. The fact that plants lack a perspective does not mean that they lack interests, cannot be injured, harmed, benefited or cured of diseases, nor alternatively that their interests are derivative interests rather than their having a good of their own. This being so, the arguments for their moral considerability (such as Routley's, Goodpaster's or mine) have good grounds to which to appeal. The more demanding sense of the phrase "good for" (meaning something like "good from the perspective of a conscious creature") needs to be carefully distinguished from its less demanding sense (meaning something like "beneficial to something with a good of its own, whether conscious or not"). So people's readiness to ask what is good for a creature, askable because of the second sense, should not be side-tracked by the requirements of the first sense, or the majority of living creatures will be disqualified from moral considerability through nothing but an equivocation. Otherwise we are driven to Glover's "mind-dependent theory of value" (see VOME, p. 21), unnecessarily and prematurely.

Problems are now raised about why the language of flourishing does not apply to species and to capitalism, entities which, like individual creatures, grow, change, face threats and either overcome them or die. However, I have explained at VOME, p.24, how we can sometimes speak of a species faring well or flourishing because we have in mind the flourishing of its current population, the members of which have the capacities needed for flourishing, and perhaps the same applies to capitalism (at a strenuous pinch). Besides, the extinction of a species deprives possible future members of existence, as well as of flourishing, and thus harms the species (concretely conceived), while damage to capitalism (again concretely conceived) similarly spells harm to coming generations unless it is replaced with a better system, which means that capitalist societies can be harmed. But construed as abstract collections or systems, neither species nor systems have either interests or the inbuilt capacities needed for flourishing. Thus we can rule them out as having moral standing without having to rule out individual creatures at the same time.

Crisp now brings to attention the perfectionist's procedural requirement to seek an independent and impartial account of a being's essential capacities, and the way in which their conception of flourishing could possibly bias (or even distort) such an account. Aristotle's argument supplies an example of such a distortion; but Aristotle's reasoning, as W.F.R. Hardie has shown, went wrong because he was looking for distinctive as opposed to essential capacities, and because he forgot about his own criterion of self-sufficiency (and thus inclusiveness) for a conception of the good, and required instead the development of whichever capacities were "highest and most god-like" (Hardie, 1968). This is how he reached his conception of the good life as one of philosophizing; his conclusion is finally reached through abandoning the search for an inclusive account of human good.

The same charge is now tried out against my own position. "It is as plausible," he writes, "to claim that human beings are characteristically inclined to rest and amusement as it is to claim that it is in their nature to engage in meaningful work". This remark is initially off target because I make no claims about human inclinations as opposed to human capacities. What the objection should rather claim is that they have essential capacities for seeking rest and amusement; and with these claims I can happily go along, without my claims about meaningful work needing to be discarded. His next example is philistinism, which I take to mean failure to develop aesthetic capacities (although it sometimes instead means exercising them in ways disdained by the utterer); but it is no problem for a perfectionist that people often fail to develop essential capacities, and to fulfil their nature thereby.

Yet why, he goes on to ask, should characteristically lazy and vulgar beings (otherwise like us), who also have capacities for accomplishments and aesthetic appreciation, be expected to fulfil their nature (presumably as lazy and vulgar people)? I can get no clear handle here on what is here being imagined, but if, on the one hand, these creatures have paired sets of conflicting tendencies in their nature, then their good would surely have to involve some blended embodiment of them; whereas if they really have essential capacities for facile art and facile judgements (rather than for significant aesthetic achievement and appreciation), as perhaps some mammalian non-human species may be held to do, then their flourishing would surely include developing the active capacities that they actually have, rather than in inactivity. However, their description makes them sound more like something different again: human beings with inbuilt capacities similar to our own, but brought up in a lazy culture; and the flourishing of people such as these would be no different from our own, the main difference lying in the need to overcome the pull of that culture on the way to such flourishing.

The following challenge is the really interesting one, grounded in the pervasive human capacities to age and to die. Here I need to make a distinction between active and passive capacities. For the capacities to age with self-respect (if one reaches a sufficient age) and to face death with fortitude are plausibly essential active human capacities, the development of which genuinely contributes to human flourishing and well-being. However, passive capacities, including essential ones, such as undergoing suffering and death, are not ones the development of which makes or could make such a contribution. For suffering and dying are not in themselves actions (any more than ailing or being injured are), and thus cannot be actively developed; indeed vulnerabilities are often distinguished from (active) capacities. Perfectionism from Aristotle onwards has been concerned with (active) capacities rather than with vulnerabilities, and that, in my view, is what modern perfectionists need to make central in their theories. I am grateful to Crisp for being pressed to present this clarification.

Beyond this point, Crisp turns to a defence of hedonism. One of the arguments presented here turns on the suggestion that the things people now value were originally valued because they were enjoyable, or because of the pleasure involved. I have already pointed out that such evolutionary narratives have little bearing on what is valuable in the present. Another problem with this approach is that it is implausible that everything was originally desired for the sake of the pleasure that it would bring, and that nothing was originally considered desirable for its own sake (not even one's health, or one's making decisions independently or for oneself). (Both Joseph Butler and David Hume made this same point long ago.) It is also highly implausible that people in the present are under an illusion when they regard accomplishment or autonomy as intrinsically desirable, and do so only because these goods were originally associated with pleasure, but have somehow become regarded as intrinsic goals by association or by mistake.

Crisp's own thought-experiment may allow us to reach a relevant judgement. We are invited to consider the contrasting cases of P, who actually writes a great novel and enjoys doing so, Q, who is on an experience machine and has P's experiences presented to her as if her own, and R, who also writes a great novel but fails to enjoy doing so. Here the hedonist, as well as claiming that if P's experience is better than Q's this is because of additional pleasures, needs to deny that what R achieves is good for her at all. One problem here is that in Nozick's experience-machine thought-experiment, Q is supposed to have exactly the same pleasures as P, and that because of this the hedonist has no reason to regard P's experiences as superior; it is not clear to me that Crisp has successfully tackled this problem. Be that as it may, I think that most people would say that R's well-being or flourishing has genuinely been furthered by R's writing the novel, despite R's complete lack of enjoyment. This could well be because developing one's capacities and fulfilling one's life-goals are valuable even in the absence of pleasure or happiness at doing so, even though pleasure and happiness enhance well-being if they are present as well.

In his penultimate paragraph, Crisp adds two further points. One is that perfectionist views tend (because of problems of perspective) to be anthropocentric, appealing to "our views about what is significant in human lives", and can even come to appear to their holders as arbitrary. But even if this is a feature of some perfectionism, it is hardly so of mine. For example, I recently contributed an address entitled 'Beyond Anthropocentrism' to the Royal Institute of Philosophy series on 'Environment' (forthcoming in the Royal Institute series volume), and have long stressed the shortcomings of anthropocentrism (in various writings) (Attfield 1991; Attfield, 1995; Attfield 2003). Besides, my stances about moral standing and about the intrinsic value of the development of essential capacities go well enough together to give this overall position some degree of stability, implying as it does an account of human well-being that includes health (whether or not the person concerned is aware of it), in parallel to its account of the flourishing of plants (without the least hint of any awareness of their part). Certainly sentientism and hedonism could also be held to go together, but sentientism does not demand hedonism (as it can instead be accompanied by, for example, belief in the intrinsic value of autonomy) in anything like the way that biocentrism (my own stance about the scope of moral standing) virtually requires the adoption of perfectionism.

The other new point of Crisp's penultimate paragraph is the supposed problem of free-will for non-hedonists. For one thing, "both libertarian and compatibilist conceptions of freedom are notoriously problematic". But they are problematic (if at all) mainly because they are each thought vulnerable to the other, in which case at least one of them would seem probably to be defensible after all. Besides, few are the defenders of hard determinism (the only remaining possibility, if libertarianism and soft determinism have to be passed over, but one that can hardly be reconciled either with moral responsibility or with the existence of capacities for choice and for self-determination). Yet Crisp seems to leave himself no other option but this one. Thus despite his view that hedonists are not obliged to provide a defence of free will, he seems committed to occupying a position far more exposed than the disjunction of compatibilism and libertarianism. (It is hard to see how any philosopher is exempted from holding one or other position from among libertarianism, soft or compatibilist determinism and hard determinism.) Fortunately there are some capable defences available for libertarianism (Wiggins, 1987; Lucas, 1970; Ward, 2004), the stance that in my view goes best with my account of the intrinsic value of autonomy and of the development of the essential human capacities for choice and self-determination.

So I am afraid that Crisp does not persuade me of the merits of hedonism or sentientism, or of departing from adherence to perfectionism and biocentric consequentialism.

References

Attfield, Robin, 1991, *The Ethics of Environmental Concern*, 2nd edn. (Athens and London: University of Georgia Press).

—. 1995, *Value, Obligation and Meta-Ethics* (Amsterdam and Atlanta, GA: Éditions Rodopi).

—. 2003, *Environmental Ethics* (Cambridge: Polity Press).

Hardie, W. F. R., 1968, 'The Final Good in Aristotle's *Ethics*', in Julius Moravcsik (ed.), *Aristotle: A Collection of Readings* (London: Macmillan), 297-322.

Lucas, J. R., 1970, *The Freedom of the Will* (Oxford: Clarendon Press).

Schloss, Jeffrey and Michael Murray (eds), 2009, *The Believing Primate* (Oxford: Oxford University Press).

Ward, Keith, 2004, 'Theistic Evolution', in William A. Dembski and Michael Ruse (eds), *Debating Design: From Darwin to DNA* (Cambridge and New York: Cambridge University Press), 261-274.

Wiggins, David, 1987, *Needs, Values, Truth* (Oxford: Blackwell).

CHAPTER TEN

ATTFIELD AND ANIMALS: CAPACITIES AND RELATIONS IN ATTFIELD'S ENVIRONMENTAL ETHICS

CLARE PALMER

Robin Attfield's work has been central to the development of environmental philosophy in a number of key areas, including stewardship, population, human development and the moral standing of living organisms. In this paper, I'll focus primarily on just one aspect of Attfield's work: human moral obligations to sentient animals. I'll first outline how, and in what ways, Attfield has argued that such animals are morally important. I'll then suggest that—while providing a good grounding for some central concerns of animal ethics—Attfield's focus on animals' capacities, in the context of a consequentialist approach to ethics, doesn't obviously accommodate other relational concerns that we might think are important. I'll consider how a consequentialist such as Attfield might respond to this concern, and I'll conclude by explaining why this still leaves some residual worries about what I'll call "capacity-oriented" consequentialist approaches to animal ethics.

Attfield's animal ethics: an outline

Attfield's ethical framework is complex and carefully developed. Here, I'll only be able to give a very basic outline of his approach to ethics in general, and to animal ethics in particular. First, and most importantly, Attfield is a consequentialist. That is, what's central on his account is to act, or to follow practices, that bring about the best possible outcomes. Best outcomes, for Attfield, are measured in terms of *flourishing*, or perhaps *well-being*, broadly understood. Attfield takes flourishing to refer not only to subjective, experiential states such as pleasure and pain, nor just to preference satisfaction or frustration (as would be standard in utilitarianism); his sense of flourishing is much more expansive. All living beings have a good of their own; they can do well or badly, they have basic needs that they must meet in order to stay alive, and in this sense they can all flourish. Admittedly, some few lives might "lack any features which make them worthwhile to anyone or anything, including the creature the life of which is in question". (Attfield 1994b, 164) But normally, all living beings can be in states of flourishing, whether or not they can consciously experience anything at all. A tree, Attfield famously argues, has a good of its own, and can flourish or otherwise; for this reason, the tree has *interests*—in receiving sufficient water, and nutrition, for instance. Non-human animals, like trees, have interests in water and nutrition, but also possess *other* interests such as not feeling pain; while humans have all these interests, plus additional interests, such as in making autonomous decisions about their own lives. Having interests and being able to flourish, then, emerges from the possession of certain *capacities*; capacities lie at the heart of Attfield's ethics. Attfield maintains: "Let the 'essential' capacities of an x be capacities in the absence of which from most members of a species that species would not be the species of x's, and let 'x' range over terms for living organisms. Then the flourishing of an x entails the development in it of the essential capacities of x's." (Attfield 1994b, 160) The development of essential capacities is what constitutes flourishing for living beings. For Attfield, ethical actions and practices are those that bring about the best outcomes in terms of developing such essential capacities by promoting and protecting organisms' interests, and thereby maximizing flourishing. Following this pattern then, in the case of animals, humans should act to (or follow practices that) maximize animals' flourishing, or, at least, maximize animals' flourishing taking into account the impact such flourishing would have on other (present or future) animal or human interests, or the interests of living but non-sentient beings such as trees.

This outline of Attfield's ethical framework is very basic—in particular with respect to animal ethics. Before moving on, I need to provide some elaborations and refinements, for Attfield carefully specifies his position, thereby pre-empting a number of potential criticisms.

First, on Attfield's account of ethics, only individual and living things can have moral standing. Although ecosystems, for instance, may be essential for the good of the living creatures that compose them, they themselves cannot be described as flourishing in a morally-relevant way. "Neither inanimate entities nor systems of living creatures have interests at all in anything other than a strained and metaphorical sense". (Attfield 1994a, 138)

Secondly, Attfield does not maintain that the flourishing of (for instance) a plant is of the same moral significance as the flourishing of a sentient animal; nor that the flourishing of a sentient animal (normally) has the same value as the flourishing of a human being. There "is much more of value in the flourishing of a sentient creature as such than in the flourishing of an individual tree as such". (Attfield 1994a, 139) Humans have a much wider range of interests than trees; promoting the flourishing of a human is more important than promoting the flourishing of a tree. As Attfield notes: "The satisfaction (or, where appropriate, non-frustration) of greater interests is of greater value than the satisfaction (or non-frustration) of lesser interests…for what is of greater value …is precisely what there is more reason for a rational agent to bring about or promote or preserve…". This doesn't mean, though, that the basic interests of an animal can be sacrificed for some relatively trivial interest of a human; an animal's interest in continuing life, for instance, is normally of more importance than the interest a human has in eating meat. The fulfillment of those interests most basic to well-being should normally have priority. (Attfield 1987, 144)

Third, Attfield's form of consequentialism is fairly closely specified. He argues for a maximizing form of consequentialism; it's the *total sum* of flourishing that should be maximized (rather than *average* flourishing). (Attfield 1991, 107) And importantly, Attfield adopts a form of *indirect* consequentialism. That is, we should not think of bringing about the best consequences primarily in terms of each individual *act*. Rather we should follow *practices* that, if they are widely followed or likely to become widely followed, would bring about best outcomes (even though in some particular case it might appear as though the best outcome would come about by *not* following the practice). So, Attfield maintains: "there are practices general recognition of which makes for, or would make for, a much better world than would be possible either in their absence or through alternative practices." (Attfield 1987, 107) So, for instance, we should keep promises, because promise-making is a practice that maximizes flourishing overall and makes for a better world; even if on some particular occasion, breaking a promise might bring about better "local" consequences, promise-breaking would undermine the socially optimific practice of promise-making.

Having given this—admittedly rather brief—outline of Attfield's ethical framework, I'll now turn to the question in which I'm particularly interested: the moral significance of capacities and relations in animal ethics.

Capacities and Relations in Attfield's Ethics

It's widely argued that moral responsibilities to others arise either on the basis of certain capacities that beings possess, or certain relations in which they stand (as Attfield himself accepts [1994c, 175]). However, Attfield's focus—as the outline of his views above suggests—is entirely on capacities, most crucially on the capacity to flourish. It's the possession of certain morally-relevant capacities that both gives a living being moral standing at all, and that provides the basis for an account of how morally significant it is, particularly important where the interests of different beings conflict. Attfield doesn't provide an account of the possible moral importance of relations at all. Indeed, he expresses a number of worries about the role that relations can play if they are given a place in ethics. For instance, he's concerned that if relations are thought to be of moral relevance, beings that lack such purportedly morally relevant relations, but that do have morally significant capacities, might be disregarded. He's also concerned that beings with "capacities of radically different types" may be treated "as if they were of equal value", on the basis of their relations, rather than their capacities. (Attfield 1991, 176) Additionally, there are some kinds of beings, he maintains, that have moral standing—such as future generations—with whom we just can't have relations (Attfield 1987, 8). And other kinds of relations—such as being someone's "kin"—are, he insists, not of moral relevance at all. (Attfield 1987, 6) Indeed, Attfield claims, "the criterion of standing in particular relationships cannot in itself be other than arbitrary". (Attfield 1991, 178) What matters for Attfield is the possession of certain capacities, capacities that ground interests and provide the basis for flourishing, not the relation in which we might stand to the beings that have these capacities.

Of course, this isn't to say that Attfield must deny that relations are *important*. After all, many relations are critical for flourishing, and should be cultivated for that reason (for instance, the relations between parent and child). But the relation is, as it were, secondary to the flourishing; it's not the relation in itself that creates particular obligations. That is, a parent should care for his or her child because such care is central to the child's flourishing and the parent is (usually) in the best position to promote their own child's flourishing, not because the relation of being someone's parent creates special caring obligations towards that child.

Attfield's view, then, is essentially *capacity-oriented*. Moral standing, and what constitutes ethical practice, is based on the flourishing that can flow from the intrinsic capacities that living beings have, independently of these beings' relations to others. This kind of capacity-oriented view is common among consequentialists, in particular in the field of animal ethics (Peter Singer and R.G. Frey, for instance, equally focus on capacities). However, it's worth noting that it's not necessarily the case that consequentialist ethical approaches *must* focus solely on capacities. Just as Attfield's ethical framework diverges from standard forms of utilitarianism by proposing a form of consequentialism in which the possession of value doesn't depend on the capacity for conscious experiences, so it would be possible to propose a form of consequentialism in which maximizing the fulfillment of certain kinds of relations carries value. I'm not going to propose such a view, though, because it doesn't obviously capture what I want to argue is morally important about certain kinds of relations. It's not that certain relations should be maximally developed or fulfilled, I'll suggest. It's rather that the existence of some kinds of special relations create some kinds of special obligations—additional responsibilities that we have towards some beings that we don't have towards others—even though those beings may have *very similar capacities*. I'll explore this idea further in the next section.

Special Relations and Obligations

Scheffler (1997, 190) identifies three possible ways in which special moral responsibilities are sometimes argued to be created by special relations. These are: (a) moral responsibilities that emerge out of past interactions (eg. promises, agreements, debts, harms); (b) moral responsibilities that emerge out of special relationships (eg. children, parents, siblings, friends) and (c) special responsibilities that emerge out of membership of some common group. I've already considered Attfield's response to several of these kinds of relations. Relations based on past interactions, such as promises, may be important on Attfield's account, but not directly because promises create special moral relations and therefore obligations. Rather, promises shouldn't be broken because this would undermine the optimific social practice of promise-making. And relations such as "being kin" are, on Attfield's view, an inadmissible basis for moral significance; the particular configuration of someone's genetic material alone is irrelevant to our moral obligations. (And after all, if we accept that just-being-kin has moral significance, it's going to be difficult to deny acceptance of less-savoury moral preferences, such as those that can be embedded in racism and sexism.)

As long as one thinks of relations such as kinship in the very "bare" sense Attfield suggests, he seems to be right. Sharing a certain sort of genetic configuration with someone else, in itself, *doesn't* seem to be a good basis for morally privileging them over others who lack this configuration. But there are other kinds of relations that, I'll argue, can hold between kin (but not only between kin) and look much more plausibly to be of moral significance. And such relations, in various forms, can extend beyond the exclusively human sphere into our moral obligations towards animals.

Let's return to parents and children, and the widely-accepted idea (in most current societies at least) that a parent has special responsibility to care for his or her own child. We've seen two possible reasons for thinking this. One is based on bare kinship; but I've accepted Attfield's view that bare kinship—genetic configuration alone—isn't morally relevant. A second reason flows from Attfield's capacity-oriented view: that we should maximize flourishing, especially human flourishing; and that a parent is usually best placed to promote their own child's flourishing. So, parents should promote the flourishing of their own children (we could think of this as one of Attfield's optimific social practices). This, of course, is plausible.

But it's not necessary to deny *this* reason to accept an *additional* reason for parental special responsibility: that the parents have decided to create, between them, a vulnerable and dependent being; that they are causally responsible for this child's existence; and that this gives them special moral obligations towards that child that they don't have towards children in general. As Onora O'Neill (1979, 26) maintains: "a standard way of acquiring obligations is to undertake them, and a standard way of undertaking parental obligations is to decide to procreate". Suppose someone decides to procreate, but denies any obligations to the infant, neglecting it or failing to provide for its basic needs. The neglectful parent is morally culpable, we normally think, in a way that would not apply to some other adult who, though knowing that there are neglected infants nearby, and being able to adopt one of them and promote its flourishing, nonetheless chooses not to do so.

Having said that we would normally think this, it's not obvious that Attfield would endorse this view, for these kinds of backward looking special obligations fit uncomfortably into a consequentialist framework. Attfield might conclude that, since the goal is to maximize flourishing, any adults should adopt needy children that they could help to flourish (provided that this did, in fact, maximize flourishing) and that such adults fail morally if they don't do so. Indeed, since Attfield (1991, 178) has argued that "the criterion of standing in particular relationships cannot be other than arbitrary", it looks as though he's likely to maintain that adults generally have just the same obligations to unrelated neglected children as parents do to their own children, even though such unrelated adults have no causal responsibility for these children's existence or their current state of well-being, provided that the unrelated adults are best placed to promote the children's flourishing. However, I want to argue, instead, that there is a relationship that generates special moral obligations in such cases, obligations that don't just flow from the child's capacity to flourish and the parents' ability to promote that flourishing.

This kind of special relationship, I suggest, combines several features. First, the child is a being that has—on the basis of capacities he or she possesses—moral status. (So, I'm not attempting to deny the highly plausible claim that particular capacities—such as sentience—are what *gives* a being moral status in the first place). Second, this relationship concerns a moral agent acting to bring a being with moral status into existence. And third, it concerns bringing a being with moral status into existence *in a particular way*—a state of vulnerability and dependence. (After all, children can be brought into the world in no other state.) It's this combination—not bare kinship, nor just the promotion of flourishing in the world—that, I suggest, gives parents special obligations to care for their children, or to arrange for other appropriate care for them.

Why am I focusing on this case? Because, I want to maintain, *some* human relations with animals are similar in form. Such human-animal relations likewise concern sentient beings with moral status that humans are largely responsible for bringing into the world, beings that are dependent on humans (indeed, have been made to *be* dependent by humans) and that are vulnerable in a variety of ways. Primarily, these are domesticated animals, although a version of this argument can be extended to include other animals rendered vulnerable by human actions, such as by displacement from their habitat.

Let's work this claim through in more detail. First, I'm following Attfield (and, of course, many others) in taking the view that sentient animals have moral status, based on particular capacities that they possess. This is relatively uncontroversial, and I'll just assume it here (one need not follow Attfield as far as the moral standing of trees to accept the moral standing of sentient animals). Second, humans are largely responsible for the actual existence of most individual domesticated animals, since they are deliberately bred. And third, domesticated animals are deliberately created in particular ways by humans that render them peculiarly vulnerable and dependent: moulding body shapes, fur or hair production, susceptibility to disease, reproductive capacity, temperament, presence or absence of horns or claws, and so on. These kinds of vulnerabilities, and the dependency that often results from them, are permanent as opposed to developmental (as is normally the case with children).

It's this deliberate causal responsibility for the creation of sentient animals in dependent states that, I suggest, means that we have special moral obligations towards them. This doesn't mean that such relational obligations hold towards all animals; for humans usually have quite different relations with wild animals; wild animals normally come into being independently of people and can provide for themselves. The argument here, then, is that while we have *some* duties towards all animals based on their possession of morally-relevant capacities, we have additional moral obligations to those animals with whom we have relevant special relations.

Admittedly, there are a number of difficulties with this argument. For instance, it might be objected that the relation of parents and children is not really parallel in form to the relations of humans to domesticated animals, since the parent/child case concerns the creation of particular individuals by other particular individuals, but most people *don't* themselves breed domesticated animals. And, of course, it's true that the parallel is not exact. However, most humans are, in various ways, *entangled with* the creation of, and certainly benefit from, the breeding of domesticated animals; and this is, I've argued elsewhere, enough to generate at least weak special moral responsibilities towards these animals. Unfortunately, I don't have space to pursue these problems further here (but see Palmer 2010 for further discussion of this, and other, objections). What I want to do now is to think about how an advocate of capacity-oriented consequentialism, such as Attfield, might respond to this argument.

Attfield and the Rejection of Special Relations/Obligations

As I've already indicated, Attfield is not obviously sympathetic to the idea that any relations are in themselves morally significant. But there are several ways in which the kinds of relations I've outlined above don't raise the problems about which Attfield actually expresses concern. First, I'm not suggesting that moral status itself is based on relations; rather that relations may create special, *additional* obligations, where beings already have moral status on the basis of their capacities. Second, a relationship that involves making a being vulnerable and dependent (whether creating the being that way, or subsequently rendering it that way) is not arbitrary in any normal sense of the term. It would be odd to call it arbitrary, for instance, were I required to compensate someone for some past harm I did to them; what's owed to this person now is directly linked to what I took from them before. The two of us are bound together in a direct historical relationship; the kinds of relations in which I'm interested are linked in a similarly non-arbitrary way.

Nonetheless, it's very difficult for a consequentialist to accept these kinds of special obligations, partly because of the forward-looking nature of consequentialism. As we've already seen, Attfield maintains that we should keep promises for forward-looking, not backward-looking reasons. Likewise, from a consequentialist point of view, the purpose of compensation or other forms of "making good" the past is to bring about the best outcome, not to engage in backward-looking reparative justice. On a forward-looking account, a human role in creating and rendering animals vulnerable and dependent is not relevant to what's owed to them. Rather, we should consider how to bring about best consequences with respect to the flourishing of animals in general, not look backwards to our causal responsibility for the present circumstances of some of them.

If this is the aim, though, the distinction I've been suggesting between what might be owed to domesticated and wild animals—or, more accurately, towards animals on whose natures or states humans have had some kind of causal impact, and animals where this isn't the case—collapses. And this is the standard consequentialist view, the view that I think should be attributed to Attfield, though he doesn't say this explicitly. We should take the flourishing of all sentient animals equally into account in our moral decision-making, whatever their history and relation to us; and aim to maximize it (taking into account the impact this would have on the flourishing of [present or future] humans, or other living but non-sentient organisms such as trees).

This consequentialist approach—at first sight at least—seems to commit those who hold it to some strong—and apparently implausible—positions. One reason for this just is the implausibility of disregarding all moral claims that arise from the past (whether or not we want to call these "justice" claims in animal cases) unless responding to such claims would anyway contribute towards the best outcome in the future. For the idea of *desert*, as James Rachels (2007) argues, is fundamentally important both to the ways we structure society and the ways in which we act in it; but desert seems to have no significant place in a wholly forward-looking ethical approach. Secondly, since the focus of consequentialism is on future states of affairs, rather than on the actors that produce such states of affairs, no distinction can be made between (for instance) states caused by actively harming, and states allowed by omitting to assist when one could have done so, leading to standard concerns about the over-demanding nature of consequentialism. Attfield (1987, 98) is explicit about this in the case of distant *people*: "The *consequences of omissions* should be recognized as including, most obviously, those states of the world (eg. avoidable starvation) which the omitted action would predictably have prevented…and the nonoccurrence of those states (eg famine relief) which the omitted action would predictably have caused." (Emphasis mine.) Attfield doesn't maintain here (unlike Thomas Pogge [2002] for instance) that the moral responsibility for relieving poverty has any relationship to one's own, or one's society's, role in causing it, or one's benefiting from it. It doesn't matter how the suffering or misery of some stranger is caused, nor how distant he or she is in time or space; if one could have prevented it, and failed to do so, one is morally responsible for it. In not acting when one could have, the total amount of flourishing in the world is less than it otherwise would have been.

But preventable distant *human* suffering is not all that's at stake here. We seem to become morally responsible, too, for distant animal suffering, if we could have prevented it: for wild animals that starve in winter storms or in droughts when we could have provided for them, or for animals with diseases against which we could vaccinate them. In not acting in the wild to increase animal flourishing when we could have, we're responsible for the fact that the total amount of flourishing in the world is less than it otherwise would have been. Yet this is surely an over-demanding view, one that both potentially commits us to constant action in the wild to promote flourishing, while having no place for the claim that we have special obligations to assist in cases of vulnerabilities we have ourselves created, in a way that we don't have towards vulnerabilities that have come about independently of us.

Arguments about implications for intervention in the wild are often leveled at capacity-oriented forms of consequentialism such as Attfield's. Of course, a capacity-oriented consequentialist might just bite the bullet on these arguments and accept that this is, indeed, an implication of the position. However, more commonly, consequentialists argue that their view doesn't necessarily commit them to negative responsibility for avoidable wild suffering; I'll briefly consider such responses below. I'll also consider whether Attfield's particular form of consequentialism—in particular his *practice*-consequentialism, rather than act-consequentialism—gives him extra tools that might provide a way of addressing these kinds of relational concerns.

Capacity-Oriented Consequentialist Responses

There are, then, two possible problems here for capacity-oriented consequentialist approaches such as Attfield's. One is that there's a strong commitment to acting in the wild to improve animal flourishing. And the other is that even where humans have been deliberately causally involved in creating animals to be dependent and vulnerable, or putting them into such states (for instance, by destroying their habitat) there's no obvious way of accommodating any special kind of moral responsibility towards them.

The first problem has attracted much more attention than the second. Consequentialists have responded to it in several ways. One possible response is to accept the implication of the view, but to maintain that, as a matter of priority, since each individual human can only do so much, and there's plenty of human and animal need around, it's unlikely that wild animal suffering will get much priority. It's likely, in practice, to be easier and more effective to relieve pain and to promote flourishing in the context of the industrial farm than the wilderness; in a world of limited resources, the farm may thus be prioritized over the forest. A second response—made by Peter Singer (1973) for instance—is to argue that we don't know the longer-term consequences of acting in the wild. What we think, in the short term, would contribute to the maximization of overall flourishing might well, in the long term, reduce it. Given our ignorance about the workings of ecosystems, it's better not to act to protect animal welfare in the wild at all.

Attfield could accept both these arguments, and combine them with his practice-consequentialism to create an even more plausible argument. After all, for an act-utilitarian, there might be any number of particular cases where neither of these concerns would hold. It's not difficult to imagine situations where assisting wild animals is relatively easy and where we've a sufficiently well-informed idea about the possible outcome of such actions, that an act-utilitarian could argue that such acts were morally required. Attfield, though, could base an argument on practices, rather than on individual acts. Generally, he could argue: following a practice of non-intervention in the wild to promote flourishing would tend to bring about the best consequences; this would create a better world than one based on any alternative practices; and it's plausible that such a practice would be adopted; so we should follow this practice. However, I'm not entirely convinced by this argument, as I'll suggest below.

In terms of the second problem—special relations to domesticated animals, or animals in other ways made vulnerable or dependent by human activities—a consequentialist can just deny that such relations are of any particular moral significance. If domesticated animals are dependent and needy, then assisting them just will promote their flourishing, and failing to do so will set their interests back; their needs are likely to be more pressing than those of wild animals that can fend for themselves; so a consequentialist would, in practice, tend to domesticated animals first. What need do we have, then, of special, backward-looking commitments? A forward-looking commitment to maximizing flourishing will still, in most cases at least, prescribe that we should take care of dependent and vulnerable domesticated animals.

Perhaps these responses—especially the proposed practice-consequentialist one—*do* suffice to settle the worry that capacity-oriented consequentialism can't accommodate important relational considerations. After all, a practice-consequentialist approach, it can be argued, would not be over-demanding, and would, in fact, deliver very similar recommendations for action as an ethical approach that attributed moral significance to such "special relations" in themselves. Yet such a conclusion leaves me residually uneasy; and in the final section I'll attempt to explain why.

Residual Uneasiness about Capacity-Oriented Consequentialism

Let's start with the question of acting in the wild to promote animal flourishing. I've suggested that a practice-consequentialist such as Attfield could argue that following the practice of non-intervention in the wild would bring about best overall consequences in terms of maximizing flourishing. But I'm not sure, on reflection, that this is obviously the case. For practice-consequentialism, in particular, is not just concerned with how individuals should act in their daily lives. It's also concerned with what practices should govern policy-making, including, for instance, what policies should govern the management of national parks, wildlife reserves and so on. In these cases, arguments that attending to the needs of starving people or domesticated animals would have priority are really irrelevant; parks will have management policies whatever the state of affairs in those other spheres might be. So, a *priorities* argument wouldn't hold. And in the last couple of decades, wildlife managers have acquired an entirely new arsenal of technology to assist them in (often literally) keeping tabs on and controlling wild animals; and there has been significant development in wild animal contraception, disease prevention and genetic modification. Given this technological advance, it's not implausible to think that wild animal flourishing in the long term may well best be achieved by intervention rather than by standing back; and after all, as Attfield insists, if there is suffering out there that we could have relieved without creating worse consequences, and we fail to relieve it, then we are morally responsible for it. Peter Singer (1973)—perhaps the best known capacity-oriented consequentialist writing about animals—himself accepts that if reducing suffering were really to result, then wildlife management would be the best strategy: "If, in some way, we could be reasonably certain that interfering with wildlife in a particular way would, in the long run, greatly reduce the amount of killing and suffering in the animal world, it would, I think, be right to interfere". It's at least an open question what practice should best be adopted as the basis for policy-making in the wild from a capacity-oriented practice-consequentialist perspective such as Attfield's.

This conclusion only generates unease, of course, if there's a concern for *other* reasons about acting to promote the flourishing of animals in the wild. Someone concerned about the protection of wildness in itself might have just such a concern, and for this reason strongly resist an interventionist conclusion. My uneasiness, however, is based on something rather weaker than this: that human duties *just don't extend* to promoting the flourishing of wild animals; except in cases where their lives have already been compromised by certain kinds of human activities (as I'll mention below) generally speaking, we should just leave them alone. This is not because by leaving them alone their flourishing is best promoted, nor because assisting them compromises their wildness or the wildness of wild places, but because what goes on with them is not our moral business. For obvious reasons, though, this view is not one to which capacity-oriented consequentialism could easily be hospitable.

A second element of my residual worry concerns those animals with whom, I've suggested, we *do* have special relations: vulnerable and dependent domesticated animals, but also (for example) wild animals that have been made vulnerable by habitat destruction. For consequentialists, the *origin* of vulnerability and dependence is irrelevant; it's current states, not past stories, on which we should focus. Yet the idea that actions in the past can create special responsibilities in the present is difficult to dismiss, and is, after all, critical to many accounts of justice. More specifically, in the animal case, if we don't look back, we can't discriminate between what we owe to (for instance) a hungry cat that we have bred and kept as a house cat, and hungry mice that have chewed their way into the house, and are scrabbling around for food as winter closes in. On a consequentialist account, we should do what promotes the most flourishing, cat and mice alike, irrespective of the ways in which we are responsible for the cat's situation and not for that of the mice. In questioning this approach, I'm not suggesting that killing such mice would be a matter of moral *indifference*, nor that there would be something morally *wrong* were we to feed them. My worry lies in the suggestion, first, that we morally *should* promote the mice's flourishing at all, and second, that our responsibility to promote the mice's flourishing is *just the same* as it is to a domestic cat for whose existence we were responsible, and whom we brought into the house and confined there.

To conclude

In this paper, I've accepted several of Attfield's key ideas: for instance that the possession of certain capacities is what gives beings moral standing, and that moral standing is possessed by individual beings rather than by species or ecosystems. However, I've also tried to argue that, alongside *capacities*, some kinds of *relations* are of moral significance in both human and animal cases, in particular, where humans are deliberately responsible for creating or inducing dependence and vulnerability in beings that have moral status. I suggested that it was very difficult for consequentialists such as Attfield to accept such relational claims, both because these claims entail looking back to the origin of particular states such as vulnerability, dependence or suffering; and because to accept such claims is to accept that factors other than bringing about the best consequences are relevant to our moral obligations. Of course, it's open to consequentialists just to reject such relational claims. But I hope that by raising them in this context, I have at least suggested some plausible difficulties with capacity-oriented consequentialist approaches to animal ethics.

References

Attfield, Robin, 1987, *A Theory of Value and Obligation* (Beckenham, Kent: Croom Helm).

—. 1991, *The Ethics of Environmental Concern*, 2nd edn. (Athens: University of Georgia Press).

—. 1994a, 'Reasoning about the Environment', in Attfield, *Environmental Philosophy: Principles and Prospects* (Aldershot: Avebury), 135-149.

—. 1994b, 'The Good of Trees', in Attfield, *Environmental Philosophy: Principles and Prospects* (Aldershot: Avebury), 153-171.

—. 1994c, 'Biocentrism, Moral Standing and Moral Significance', in Attfield, *Environmental Philosophy: Principles and Prospects* (Aldershot: Avebury), 173-182.

Frey, R.G., 2007, 'Moral Standing, the Value of Life and Speciesism', in *Ethics in Practice*, 3rd edn., edited by Hugh Lafollette (Oxford: Blackwell), 192-204.

O'Neill, Onora, 1979, 'Begetting, Bearing and Rearing', in *Having Children: Philosophical and Legal Reflections on Parenthood,* edited by Onora O'Neill and William Ruddick (New York: Oxford University Press), 25-38.

Palmer, Clare, 2010, *Animal Ethics in Context* (New York: Columbia University Press).

Pogge, Thomas, 2002, *World Poverty and Human Rights* (London: Polity Press).

Rachels, James, 2007, 'Punishment and Desert', in *Ethics in Practice*, 3rd edn., edited by Hugh Lafollette (Oxford: Blackwell), 510-518.

Scheffler, Samuel, 1997, 'Relationships and Responsibilities', *Philosophy and Public Affairs*, 26, 189-209.

Singer, Peter, 1973, 'Food for Thought', *New York Review of Books* 20/10 June 14th, http://www.nybooks.com/articles/9822 (Accessed November 13th 2009).

CHAPTER ELEVEN

REPLY TO CLARE PALMER

ROBIN ATTFIELD

Clare Palmer presents a sympathetic and fair-minded account of my theory of biocentric practice-consequentialism, and conveys an almost entirely accurate interpretation of it. However, my prioritising of basic interests over relatively trivial ones is actually paired with a second criterion of moral significance, that of the psychological complexity of the creature's capacities (Attfield, 1983 and 1991: 173-7; 1995: 92), and because of this pairing of two criteria, both relevant to inter-species ethics, I am not committed to the wrongness of eating meat as such, although I accept the wrongness of "consuming the products of factory-farms and of other practices which cause significant animal suffering without sufficient reason" (1983 and 1991: 181).

To turn to Palmer's narrative of my account of capacities and relations, I do indeed hold that "the criterion of standing in particular relationships cannot in itself be other than arbitrary" (Attfield, 1983 and 1991:178), but the emphasis here is on "in itself". As she says, I can still hold that relations are important (albeit secondary). And because of this I have no need to deny (as she supposes) that "the relation of being someone's parent creates special caring obligations towards [their] child". For consequentialists such as myself can recognise as overall beneficial (or optimific) practices like family life, which generate special obligations for family members such as parents (Attfield, 1995: 108-113); and so, given this practice, becoming a parent can (and in my view does) create special caring obligations. This theme of how consequentialism upholds relationships and related obligations is ably elaborated by Peter Railton in 'Alienation, Consequentialism and the Demands of Morality' (Railton, 1988); as Railton argues, there is no need for consequentialism to be understood in a way that alienates its adherents from their relationships, as might well be the case if our love of family members had to be provisionally given on a day-to-day basis on the condition of it continuing to serve the general good. (Railton subscribes to a different kind of consequentialism from mine, but his reasoning arguably remains relevant; it would take us too far afield to argue the point here.)

For similar reasons, when Palmer claims that "relations such as 'being kin' are, on Attfield's view, an inadmissable basis for moral significance" and thus represents me as holding that "the particular configuration of someone's genetic material" (which may, of course, correlate with being a parent, child, uncle, aunt or cousin) is morally irrelevant, I need to qualify this interpretation; for, while being someone's kin is not the ultimate basis of obligation, the family system confers on it moral significance which is often profound. Thus I do not hold that "adults generally have just the same obligations to unrelated neglected children as parents do to their own children" (for parents have special obligations as parents), although if these adults are "best placed to promote [these neglected children's] flourishing", I do consider (with Peter Singer) that they are not without obligations in the matter.

However, it is not clear that procreating or bearing a child is invariably what confers such special obligations. For adopting a child confers special obligations that are equally strong, and has an equally powerful consequentialist underpinning, even though the adoptive parents are not directly responsible for the existence or the vulnerability of the child. (My obligations to my adopted son can hardly be weaker than those to my two daughters whose natural father I am.) Besides, there must be cases where the natural parents are or become unable to undertake the special obligations to their children characteristic of parents, or even, as Palmer suggests for such cases, to arrange for other appropriate care. Thus while generally endorsing the special obligations of parents as depicted by Palmer, my understanding of the reasons is rather different, and does not in all cases involve bringing vulnerable creatures into being.

But that is no reason for denying that human beings usually have special obligations towards the domesticated animals in their charge. While (*pace* Palmer) not all domesticated animals are "peculiarly vulnerable", she is right in holding that their hereditary make-up has often been moulded by human beings; and this characteristic belongs all the more to genetically modified creatures. However, the key difference between domesticated and wild animals is that they are (or have been) subject to human charge, and that obligations arising from this dependency attach to their current custodians. (Much the same applies to non-domesticated animals held in human custody.) These obligations too arise, on a consequentialist view, from the practices of farming and animal-rearing, which are often (although with widespread and pronounced exceptions) optimific. (Some domesticated animals, of course, are treated as family members, and become peripheral members of the practice of family life as well.) So I do not deny that we have moral obligations to those animals with which we have relevant special relations, additional to those that we have towards all animals, based on their capacities and vulnerabilites.

Far from being inclined to deny this, I attempted to affirm the relevant principles in my 1989 review of Holmes Rolston's book *Environmental Ethics*. Rolston had defended separate principles for the treatment of sentient beings, depending on whether they were to be found in the realm of culture or that of (wild) nature. My response, in line with what Palmer now maintains, is that the underlying obligations apply to both realms (as Singer's principle of equal consideration for equal interests would suggest), but that usually the case for non-intervention in the realm of wild nature overrides the case for intervention to prevent suffering there. Thus the case for averting the suffering of domesticated animals is usually stronger, since this is the realm of human control and animal dependence on human custodians. Implicitly I was endorsing the case for special obligations towards domesticated animals (Attfield, 1989: 363-4). The bearing of these principles on factory farming was developed in *Value, Obligation and Meta-Ethics* (Attfield, 1995: 91). Thus Palmer's subsequent suggestion that for practice-consequentialists such as myself the distinction between obligations to domesticated and to wild animals "collapses" is wide of the mark.

In her next sub-section, Palmer introduces another example of special relations, that of someone owing compensation to another for some past harm done to them; these relations too may create special obligations. She adds that it is very difficult for a consequentialist to accept special obligations of these kinds, because the reasons are backward-looking, whereas consequentialism is forward-looking. However, the practice of reparations is clearly an optimific one, and this is what in my view justifies its backward-looking requirements. I argued this case a long time ago in 'Unto the Third and Fourth Generation' (Attfield, 1979b: 55-70); the title concerns the long sequence of generations over which the rectification of past wrongs may be due. I also pointed out that such duties will sometimes conflict with, and could be overridden by, obligations either of alleviating present suffering or of averting injustice in the present. At other times, however, all three kinds of obligation are liable to support policies such as overseas aid, particularly to former colonial territories. In any case the backward-looking nature of obligations to compensate for past injuries cannot be regarded as any more of a problem for consequentialism than the backward-looking obligations of other optimific practices, such as (some kinds of) punishing and (some kinds of) rewarding. (Kinds that are not optimific will not of course be endorsed by a practice-consequentialist at all.) Further, since punishing and rewarding turn in large part on past deserts, issues of desert need not be disregarded or minimized by practice-consequentialists (as Palmer seems to think). Another optimific practice which turns on backward-looking considerations, as Palmer recognizes, is that of promising, and here practice-consequentialists are free to take as seriously as anyone else issues such as whether particular past words or writings amounted to a promise and committed the utterer or writer to obligations strong enough to override other ethical factors in the present.

Palmer rightly draws to attention that, for consequentialists, agents are equally responsible for the impacts of their actions and of their omissions, although it is worth adding that, from the same perspective, they are not responsible for either the actions or the omissions that they could not have helped, and that responsibility for omissions is at least reduced when we are unaware of them or their impacts, just as it is in the case of what we do or bring about unknowingly; with actions and omissions, we must compare like with like, as I once argued in *Mind* (Attfield, 1979a). It is, however, inaccurate to say that, for consequentialists, the moral responsibility for relieving poverty has no "relationship to one's own, or one's society's, role in causing it, or one's benefiting from it"; as we have seen, practice-consequentialists uphold practices like reparations, and comparable practices such as punishment, to which these backward-looking causal factors are crucial. Rather than ignoring causal factors, consequentialists affirm the causal role of actions, and of omissions too, and this will affect their view of when (say) reparations are due, which will include cases of people inactively but avoidably benefiting from poverty and the systems that cause it as well as from making it happen. Indeed it is wrong to claim that, for consequentialism, "It doesn't matter how the suffering or misery of some stranger is caused…; if one could have prevented it, and failed to do so, one is morally responsible for it." Besides confusing moral obligation and moral culpability, as if my having an obligation to prevent something meant that I am morally responsible for its existence, this passage ignores the exculpatory role of not knowing what we are either causing or failing to prevent, as in the case of our ignorance of greenhouse gas emissions before the mid-1980s.

Importantly, several of the practices upheld by practice-consequentialism (and outlined above) take seriously people's obligations to avoid omissions where either caring or preventative action was needed, and thus condemn neglect or negligence as well as the active causing of harm.

Family duties are clearly a case in point, as are the duties attaching to the rearing of animals. Neglect can also (even on ordinary interpretations) trigger expectations of restitution, and within systems of criminal justice can justify censure and punishment. Practice-consequentialists, given their beliefs about the parity of actions and comparable omissions, are thus consistent in supporting practices that focus on the impacts of inaction as well as those of action; and their emphasis on the impacts of inaction means that their stress on the causation of suffering and misery is more comprehensive than the view that rejects what Bernard Williams called "negative responsibility" (Williams, 1988) and holds that we have no responsibility for what we let happen when we could have prevented it. Yet their doing so need not make their stance an over-demanding one (another problem mentioned by Palmer); in this matter, Railton supplies some excellent replies to Williams, which are well supplemented by those of Paul Gomberg (1986), and need not be rehearsed here.

Palmer proceeds to suggest that the consequentialist recognition of negative responsibility implies an obligation for constant action to prevent animal suffering and promote flourishing in wild nature; if we do not accept this role, "the total amount of flourishing in the world" will be "less than it otherwise would have been". This, however, presupposes that human intervention would be well enough informed as to be likely to produce an overall gain to flourishing, or an overall reduction of suffering. But by and large human beings (as I have suggested above) almost certainly lack the understanding that would be needed, and if so, the state of the world that "would otherwise have been" would as often as not be worse than if we leave ecosystems alone. There are certainly exceptions, such as cases where animals are about to die of thirst at a shriveling water-hole, and humans could move them to a suitable environment outside the area of drought; here, inaction would probably make the world worse than "it otherwise would have been". Another kind of case might concern a species, driven by global warming to the poleward extremity of the territory to which it can migrate; in some cases where human agency could move the species to another viable habitat, it could be right or even a responsibility to do so. (Both these kinds of action would also probably be supported by the precautionary principle, which urges intervention to prevent ecological catastrophes, and which consequentialists can consistently endorse.) But the generic interference and policing of nature that Palmer seems to impute to consequentialists (or at least to act-consequentialists) distorts what they could ever be expected to undertake; consequentialists would be no better served by arrogance than anyone else. Indeed Palmer recognises that a practice-consequentialist such as myself could well favour a practice of non-intervention in the wild, since it could be held both optimific and likely to be widely adopted.

But Palmer finds this kind of practice-consequentialist reply unsatisfactory. She rightly rejects the view that normative theories relate to individuals only, who have too little time left for wild animals after campaigning about the treatment of domesticated ones, for such theories relate also to the policies of governments and of park managements. However, park managements would be subject to the same constraints as those mentioned above, whatever the technology that is now available to them by way of animal contraception and disease prevention.

Yet her basic reason is a conviction that what goes on in wild nature is not our business, except where humanity has generated the problems. But here I simply disagree; we should, in my view, intervene in the case of the shriveling water-hole whether the drought is partly caused by anthropogenic global warming or entirely by natural causes. As for her cat and mouse example, I have explained how I can recognise a special responsibility to the cat, who has been taken on as a member of the household; but I disagree if her view is that we are free to be entirely unconcerned about the mouse, since I am clear that we should not allow it to be tortured when we could prevent this, even if we can only prevent this by killing it.

Accordingly I see no reason to introduce relations as a distinct basis for moral responsibilities alongside capacities. There are genuine responsibilities associated with relations, but these responsibilities derive from optimific practices which involve both the past and/or relations, such as family life, animal rearing, reparations, reward and punishment, and promise keeping. Consequentialists need not "reject such relational claims", but need not regard relations as ultimate justifications either.

References

Attfield, Robin, 1979a, 'Supererogation and Double Standards', *Mind*, LXXXVIII, 481-99.

—. 1979b, 'Unto the Third and Fourth Generation', *Second Order*, VIII.1&2, 55-70.

—. 1983, *The Ethics of Environmental Concern*, 1st edn. (Oxford: Basil Blackwell, and New York: Columbia University Press).

—. Review of Holmes Rolston, 1988, *Environmental Ethics: Duties to and Values in the Natural World* (Philadelphia: Temple University Press), in *Environmental Ethics*, 11.4, Winter 1989, 363-68.

—. 1991, *The Ethics of Environmental Concern*, 2nd edn. (Athens, GA and London: University of Georgia Press).

—. 1995, *Value, Obligation and Meta-Ethics* (Amsterdam and Atlanta: Rodopi).

Gomberg, Paul, 1986, 'Self and Others in Bentham and Sidgwick', *History of Philosophy Quarterly*, 3, 437-448.

Railton, Peter, 1988, 'Alienation, Consequentialism and the Demands of Morality' in Samuel Scheffler (ed.), *Consequentialism and Its Critics*, (Oxford and New York: Oxford University Press), 93-133.

Rolston, Holmes, III., 1988, *Environmental Ethics: Duties to and Values in the Natural World* (Philadelphia: Temple University Press).

Williams, Bernard, 1988, 'Consequentialism and Integrity', in Samuel Scheffler (ed.), *Consequentialism and Its Critics* (Oxford and New York: Oxford University Press), 20-50.

WORKS OF ROBIN ATTFIELD

Since this is a book about the works of Robin Attfield it is appropriate that a full list of his publications is presented here. Robin Attfield provides a brief description of each of the articles and chapters.

Books

1. Robin Attfield, *God and The Secular: A Philosophical Assessment of Secular Reasoning from Bacon to Kant*, Cardiff: University College Cardiff Press, 1978, pp. 231, ISBN 90142692-X.
2. Robin Attfield, *The Ethics of Environmental Concern*, Oxford: Basil Blackwell Publisher, 1983, pp. xi + 220, ISBN 0-631-13137-X, and New York: Columbia University Press, ISBN 0-231-05798-9 (hb), 0-231-05799-7 (pb).
3. Robin Attfield, *A Theory of Value and Obligation,* London, Sydney, New York: Croom Helm, 1987, pp. x + 262, ISBN 0-7099-0572-6.
4. Robin Attfield and Katharine Dell (eds.), *Values, Conflict and the Environment* (Report of the environmental ethics interdisciplinary Working Party of the Ian Ramsey Centre, St. Cross College, Oxford), Oxford: Ian Ramsey Centre and Cardiff Centre for Applied Ethics, 1989, pp. 100, ISBN 1-873453-00-0.
5. Robin Attfield, *The Ethics of Environmental Concern* (2nd edn.), Athens, GA and London: University of Georgia Press, 1991, pp. xxiii + 249, ISBN 0-8203-1349-1 (hb), 0-8203-1344-0 (pb).
6. Robin Attfield and Barry Wilkins (eds.), *International Justice and the Third World: Essays in the Philosophy of Development*, London and New York: Routledge, 1992, pp. ix + 207, ISBN 0-415-06924-6 (hb); 0-415-06925-4 (pb).
7. Robin Attfield, *God and The Secular: A Philosophical Assessment of Secular Reasoning from Bacon to Kant* (2nd edn.), Aldershot: Gregg Revivals and Brookfield, VT: Ashgate, 1993, pp. 231, ISBN 0-7512-0243-6.
8. Robin Attfield, *Environmental Philosophy: Principles and Prospects*, Aldershot: Avebury and Brookfield, VT: Ashgate, 1994, pp.viii + 262; ISBN 1-85628-566-9.

9. Robin Attfield and Andrew Belsey (eds.), *Philosophy and the Natural Environment* (also published as Royal Institute of Philosophy Supplement 36), Cambridge: Cambridge University Press, 1994, pp. vi + 250, ISBN 0-521-46903-1.
10. Robin Attfield, *Value, Obligation and Meta-Ethics*, Amsterdam and Atlanta: Rodopi, 1995, pp. xv + 319; ISBN 90-5183-857-3 (hb); 90-5183-862-X (pb).
11. Robin Attfield and Katharine Dell (eds.), *Values, Conflict and the Environment* (Report of the environmental ethics interdisciplinary Working Party of the Ian Ramsey Centre, St. Cross College, Oxford), 2nd edn., Aldershot: Avebury, and Brookfield, VT: Ashgate, 1996, pp. xii + 174; ISBN 1-85972-491-4.
12. Robin Attfield, *The Ethics of Environmental Concern*, (2nd. edn.), translated into Korean by Seunghoe Koo, Seoul: Earth Love Publications, 1997, pp. 377, ISBN 89-85277-21-9.
13. Robin Attfield, *The Ethics of the Global Environment*, Edinburgh: Edinburgh University Press, 1999, in the World Ethics Series edited by Nigel Dower, pp. viii + 232, ISBN: 07486-0895-8; also West Lafayette, IN: Purdue University Press, 1999, ISBN: 1-55753-189-7.
14. Robin Attfield, *Environmental Ethics: An Overview for the Twenty-First Century*, Polity Press, Cambridge, and Blackwell, Malden, MA, 2003, pp. xii + 232. ISBN: 0-7456-2737-4 (hb); 0-7456-2738-2 (pb).
15. Robin Attfield, *Creation, Evolution and Meaning*, Aldershot, UK and Burlington, VT: Ashgate, 2006, pp. ix + 234. ISBN 0-7546-0474-8 and 0-7546-0475-6.
16. Robin Attfield (ed.), *The Ethics of the Environment*, Farnham, UK and Burlington, VT: Ashgate, 2008, pp. xxx + 620, ISBN, 978-0-7546-2786-9.

Articles or Chapters in Academic Journals or Books

1. 'Berkeley and Imagination', *Philosophy*, 45, July 1970, 237-39. Berkeley did not run together the distinctions between reality and imagination, reality and dreams and reality and illusion.
2. 'Non-Tentative Religious Beliefs and Rationality', *Sophia*, IX, July 1970, 16-21. A vindication of the rationality of religious belief as depicted by Professor Basil Mitchell.
3. 'Ffydd wedi ei Hail-wampio ar gyfer Athroniaeth' ('Faith Refashioned to Fit Philosophy'), Review Article of D. Z. Phillips, *Faith and Philosophical Enquiry*, London: Routledge and Kegan Paul, 1970;

translated into Welsh by Canon O. G. Rees, in *Diwinyddiaeth*, XXII, 1971, 26-29. The philosophy of religion of D. Z. Phillips is criticised for its disdain of Natural Theology.

4. 'Talents, Abilities and Virtues', *Philosophy*, 46, July 1971, 255-58. An assessment of Hume's arguments and motives for declaring voluntariness inessential to virtue.
5. 'The Individuality of God', *Sophia*, X, April 1971, 20-27. Against Michael Durrant it is argued that, 'God' being sometimes a proper name and sometimes a common name, God is an individual of some not wholly incomprehensible sort.
6. 'Theology in the Modern University', in *Education for Development*, 1.3, March 1971, 23-29. A new look at the logic of the Religious Studies curriculum and the place of philosophy within it.
7. 'Collective Responsibility', in *Analysis*, 32.1, October 1971, 31-32. Statements about the responsibility of Governments are not reducible to statements about the responsibility of Cabinet Ministers alone, even though some statements of the latter sort must be true for any of the former sort to be true.
8. 'On Translating Myth', in *International Journal for Philosophy of Religion*, 11.4, Winter 1971, 228-45. A theory of how to translate passages expressing concepts once significant but now obsolete is developed and applied to a Biblical example.
9. 'Belief in God', *Sophia*, July 1972, XI.2, 1-4. The theistic belief that God is good is not as obviously absurd as J.J. McIntosh thinks.
10. 'On Euthanasia', in *Contact* (Journal of The Scottish Pastoral Association, The Clinical Theology Association and The Institute of Religion and Medicine), 37, March 1972, 10-15. Argues for the moral acceptability of voluntary contractual euthanasia.
11. 'The Irreducibility of "Meaning"' (written jointly with Mr. Michael Durrant), in *Noûs*, VII.3, September 1973, 282-98. Demonstrates the circularity of 'use' theories of meaning and of analyses of meaning of any other sort.
12. 'The God of Religion and the God of Philosophy', in *Religious Studies*, 9, March 1973, 1-9. Discusses the logical relations between natural theology and the religious appeal to revelation: concludes that we always need reasons for regarding a passage as revelatory.
13. 'How Things Exist: A Difficulty' in *Analysis*, 33.4, March 1973, 141-43. A criticism of Professor G. Owen's analysis of 'existence'.

14. 'Dr. Azikiwe's Fourth Arm', in *Second Order*, 11.2, July 1973, 86-96. A critical survey of some arguments for retaining the military as an arm of the government of Nigeria.
15. 'Did I Go To Canterbury Willy Nilly?' in a Royal Institute of Philosophy collection, *Philosophy and Psychology*, edited S.C. Brown, Macmillan, 1974, 336-38. A note on determinism. (Published without its title.)
16. 'An Anomaly of Anomalous Monism', in the same collection (as 15 above, 1974), 62-63. A note on materialism. (Published without its title.)
17. 'The Logical Status of Moral Utterances' in *The Journal of Critical Analysis*, IV.2, July 1972 (published 1974), 70-84. It is argued that even basic moral propositions have grounds.
18. 'On Being Human' in *Inquiry*, 17, 1974, 175-92. Argues that the development of certain capacities is necessary for living well as a human.
19. 'Against Incomparabilism' in *Philosophy*, 50, 1975, 230-34. Argues against the view that diverse moralities cannot morally be compared.
20. 'Towards a Defence of Teleology' in *Ethics*, 85.2, January 1975, 123-35. Contends that rule-teleology supplies a criterion of moral rightness adequate to the requirements of social justice and preferable to rival theories.
21. 'Philosophy and Secularisation', in *Thought and Practice* (Journal of the Philosophical Association of Kenya), 2.2, 1975, 135-48. A critique of secularisation over science, ethics and natural theology.
22. 'Clarke, Collins and Compounds' in *The Journal of the History of Philosophy*, XV.1, 1977, 45-54. A critical discussion of a historical debate about reductionism.
23. 'The Lord is God: There is No Other' in *Religious Studies*, 13, January 1977, 73-84. A detailed reply to Michael Durrant's book *The Logical Status of 'God'*.
24. 'Racialism, Justice and Teleology' in *Ethics*, 87.2, January 1977, 186-88. A reply to a critic of item 20 (above).
25. 'Science and Creation' in *The Journal of Religion*, 58.1, January 1978, 37-47. The paradoxical relations between science and theism are explored.
26. 'Supererogation and Double Standards', *Mind*, LXXXVIII, 1979, 481-99. Distinct but related standards of the morally desirable and the morally obligatory are expounded and defended, and with them a theory of supererogation.

27. ‘How not to be a Moral Relativist’, *The Monist*, 62.4, 1979, 510-23. Recent defences of moral relativism are found to escape self-contradiction only at the expense of incurring the charge of implausibility, and to explain nothing which could not be better explained without resorting to them.
28. ‘Unto the Third and Fourth Generation’, *Second Order*, VIII.1&2, 1979, 55-70. A theory is developed of reparations due from and to collections of people lasting more than one generation, and is related to a teleological theory of moral rightness.
29. ‘The Value of Value Inquiry’, *Lettera*, 19, 1980, 113-22. Value inquiry is defended against the widespread belief that values are subjective and arbitrary. Inquiries in ethics, aesthetics and politics into what is of value can disclose important truths.
30. ‘Religious Symbols and the Voyage of Analogy’, *International Journal for Philosophy of Religion*, XI.4, 1980, 225-38. The traditional theory of analogy makes sense of religious language only if one non-analogical proposition about God can be identified. This is supplied by his nature as potential creator.
31. ‘Woolhouse’s Open Future: Not in the Script’, *Studia Leibnitiana* (Journal of the Gottfried-Wilhelm-Leibniz-Gesellschaft, Hanover), 12.2, 1980, 229-35. An interpretation of Leibniz’s Principle of Pre-Determinate History, on which the future of substances is partially an open one, is criticised as lacking support and irreconcilable with the texts concerned, and a more traditional interpretation, on which the careers of substances cannot fail to unfold in accordance with their complete concepts, is preferred.
32. ‘How Not to Undermine Theology’, *New Blackfriars*, 61, 1980, 286-92. Far from the assertion that the divine nature is partially knowable shaking the foundations of the doctrine of creation, that doctrine is subverted rather by the negation of that assertion.
33. ‘Plantinga and Leibniz’, a critical study of Alvin Plantinga, *The Nature of Necessity*, in *Studia Leibnitiana*, 12.2, 1980, 215-20. Four related issues where the views of Plantinga have a bearing on those of Leibniz are discussed: essentialism, trans-world identity, the ontological argument and the issue of whether God can create any possible world.
34. ‘The Good of Trees’, *Journal of Value Inquiry*, 15, 1981, 35-54. Trees not only have needs of their own, *pace* Professors Hare and Feinberg, but are also of some slight intrinsic value.

35. 'Prescriptivity and Justification', (written jointly with Michael Durrant), *Philosophical Papers*, X, 1981, 16-23. Against D.Z. Phillips and H.O. Mounce, it is contended that moral rules as well as moral judgements are susceptible of justification and appraisal.
36. 'Thomas Traherne and Intrinsic Value', in Edgar Morscher and Rudolf Stranzinger (eds.), *Proceedings of the Fifth International Wittgenstein Symposium*, Vienna: Verlag Hölder-Pichler-Tempsky, 1981, 97-99. The poems and meditations of Traherne supply grounds independent of theology and ethics for the identification of both instrumental and intrinsic value.
37. 'Science, Christianity and the Roots of Our Ecological Problems', in *Lettera*, seconda serie, I, 1982, 35-47, translated by Dr. S. Gamberini as 'Teoria scientifica, etica cristiana e le origini del problema ecologico'. Christianity is largely responsible for the rise of science and technology, but is not, *pace* Lynn White, a source of ecological disruption, and has advocated an ethic of stewardship for longer than Passmore recognises.
38. 'Optimific, Right but not Obligatory', *Canadian Journal of Philosophy*, 12, 1982, 317-20. Utilitarianism can be defended against problems about supererogation raised by McConnell in *Ratio* without incurring the objections attracted by the account of Sikora in *Canadian Journal of Philosophy*.
39. 'Christian Attitudes to Nature', *Journal of the History of Ideas*, 44, 1983, 369-86. Christian attitudes to nature have not been characteristically despotic. Biblical, patristic, medieval and modern Christian attitudes have been much more varied, and overall much more benign, than their traducers contend.
40. 'Western Traditions and Environmental Ethics', in Robert Elliot and Arran Gare (eds.), *Environmental Philosophy: A Collection of Readings*, St Lucia: University of Queensland Press, 1983, 201-30. Historically the tradition of stewardship has been much more prominent than Peter Singer suggests; accordingly the roots of an environmental ethic exist already, and there is no need for a new ethic to be devised, even if this were possible. This work was also published in 1983 at Milton Keynes by The Open University Press, and at University Park, Pennsylvania by Pennsylvania State University Press.
41. 'The Threat of Wrong', Review Article of Jeff McMahan, *British Nuclear Weapons, For and Against*, London: Junction Books, 1981; J.E. Hare and Carey B. Joynt, *Ethics and International Affairs*, London and Basingstoke: Macmillan Press, 1982; Geoffrey Goodwin

(ed.), *Ethics and Nuclear Deterrence*, London and Canberra: Croom Helm, 1982; and Nicholas Sims (ed.), *Explorations in Ethics and International Relations, Essays in Honour of Sydney D. Bailey*, London: Croom Helm, 1981, in *Review of International Studies*, 9, 1983, 147-52. Though the conditional intention to use nuclear weapons is not wrong in itself, the actual consequences of current postures, compared with those of moderate unilateral disarmament, show the latter to be right.

42. 'Unilateralism, Morality and Faith', *Scottish Journal of Religious Studies*, 4, 1983, 3-14. Annette Baier is correct in holding that there are secular reasons for unilateral acts of justice in an unjust world, but they are not confined to the prospect of a Kantian kingdom of ends. They (like Christian ethics) include consequentialist considerations. But there are additional religious reasons for unilateralism.
43. 'Methods of Ecological Ethics', *Metaphilosophy*, 14, 3&4, 1983, 195-208. Ecological ethics needs to rely upon a method of argument by analogy, starting from received judgements, rather than appealing to loyalty to the planetary biosphere.
44. 'Necessity and Contingency in God', Review Article of Keith Ward, *Rational Theology and the Creativity of God*, Oxford: Basil Blackwell Publisher, 1982, in *New Blackfriars*, 64, 1983, 35-41. Ward exaggerates both God's necessity and his contingency; his proper project of combining them in a single concept thus fails in its execution.
45. 'Fairness to the Third World', *Crucible* (journal of the Church of England General Synod's Board of Social Responsibility), October to December 1983, 171-76. Effort to assist the economic and social development of the Third World is held to be obligatory on grounds not only of reparations but also of international justice.
46. 'Miller, Kripke, Bach and the Meaning of Proper Names', *Southern Journal of Philosophy*, 21, 1983, 153-58. Theories of proper names which deny their equivalence with non-trivial descriptions (which supposedly would fail to apply to the bearers of the names in some possible worlds) are confronted with counterexamples which are argued either to invalidate the theories, or at least to narrow the scope of their application.
47. 'Population Policies and the Value of People', *Journal of Social Philosophy*, 14.4, 1983 (special conference edition), 84-93. A modified version of the Total Theory of utilitarianism is defended as

the soundest basis for policies of curtailing population growth which at the same time recognise the value of people.

48. 'Thomas Traherne and the Location of Intrinsic Value', *Religious Traditions*, 6, 1983, 66-74. (The original of which 36 above is a summary.)
49. 'Work and the Human Essence', *Journal of Applied Philosophy*, 1, 1984, 141-50. Meaningful work is argued to be necessarily good for human beings, granted their nature and need for self-respect. As paid work supplies many with their only prospect of meaningful work, the opportunity of employment should be open to everyone.
50. 'Value in the Wilderness', Review Article of Donald Scherer and Thomas Attig (eds.), *Ethics and the Environment*, Englewood Cliffs: Prentice Hall, 1983, in *Metaphilosophy*, 15, 1984, 289-304. The contributors dealing with applied issues neglect key theoretical considerations raised by those dealing with fundamental issues; the valuable work of the latter is sifted, and suggestions are made about how to fill the gaps which it leaves.
51. 'Metodi di etica ecologica', *Lettera*, III serie, 5, July 1985, 3-23. A translation into Italian by Professor S. Gamberini of 43 (above).
52. 'Balthasar Bekker and the Decline of the Witch-Craze: The Old Demonology and the New Philosophy', *Annals of Science*, 42, 1985, 383-395. Balthasar Bekker's Cartesian and theological criticisms of witch-beliefs were more radical and significant than Hugh Trevor-Roper allows, and help to fill a gap in the explanation of the decline of these beliefs recognised by Keith Thomas.
53. 'The Ethics of Environmental Concern: A Reply to David H. Bennett', *The Deep Ecologist*, 17, July 1985, 11-13. A response to some criticisms of the treatment of Deep Ecology in *The Ethics of Environmental Concern*.
54. 'Development: Some Areas of Consensus', *Journal of Social Philosophy*, 17.2, Summer 1986, 36-44. Common ground can be found among development theorists both about the concept of development, and about the causes and remedies of underdevelopment.
55. 'The Prospects for Preservation', *Philosophical Inquiry* (New York), 8, 1986 (special issue on Ecology and Philosophy). Some divergences and convergences of view with Janna L. Thompson's 'Preservation of Wilderness and the Good Life' are explored.
56. 'The Good of Trees' (34 above), republished in Donald VanDeVeer and Christine Pierce (eds.), *People, Penguins and Plastic Trees, Basic*

Issues in Environmental Ethics, Belmont, California: Wadsworth Publishing Company, 1986, 96-105.

57. 'Teaching Environmental Philosophy at a British University', *Journal of Environmental Education* (New York), 18.4, Summer 1987, 15-18. The introduction of environmental philosophy into four philosophical courses at University College Cardiff is discussed, Philosophical Aspects of Social and World Problems, Philosophical Theology, Moral Philosophy and Modern European Ideas.
58. 'Biocentrism, Moral Standing and Moral Significance', *Philosophica* (Ghent, Belgium), 39, 1987, 47-58. A critique of Paul Taylor's biocentric account of moral significance, incorporating arguments in favour of a biocentric account of the scope of moral standing.
59. 'The Ethics of Power', *African Philosophical Inquiry*, 1.2, July 1987, 141-156. A theory is expounded and defended of the justification, scope and limits of the activities of collective bodies such as the state, and of the individual's obligation to obey.
60. Three contributions on the concept of person, in Arthur Peacocke and Grant Gillett (eds.), *Persons and Personality*, Oxford and New York: Basil Blackwell, 1987, pp. 92, 193, 195f. Brief criticisms are advanced of Derek Parfit on teletransportation, and of orthodox theologians on the doctrine of the incarnation.
61. 'Overpopulation', *Proceedings of the Seventeenth World Congress of Philosophy* (Montreal, 1983), Montreal: Editions Montmorency, 1988, Vol. V, 554-555. The assumptions of believers in overpopulation are criticised, and the value of nonhuman lives is related to the issue of human numbers.
62. 'Population Policies and the Value of People' (a revised version of 47 above), in Yeager Hudson and Creighton Peden (eds.), *Philosophical Essays on the Ideals of a Good Life*, Lewiston, New York: The Edwin Mellen Press, 1988, 191-201.
63. 'Intercultural Values and Human Needs', *Proceedings of the Seventeenth World Congress of Philosophy* (Montreal, 1983), Editions Montmorency, 1988, Vol.II, 491-97. Even those who deny that rational comparisons are possible between different societies or different ideological discourses are argued to presuppose concepts of human beings and human needs which supply a basis for such comparisons.
64. 'Methods of Ecological Ethics' (43 above), republished with an updating postscript in Terrell Ward Bynum and William Vitek (eds.) *Applying Philosophy* (Metaphilosophy Monographs Series), Oxford

and New York: The Metaphilosophy Foundation in association with Basil Blackwell, 1988, 176-190.

65. ‘Some Current Values and Assumptions’, jointly composed with two other members (Dr. Arthur Peacocke and Dr. Katharine Dell) of the Ian Ramsey Centre, St. Cross College, Oxford Working Party, and comprising Chapter 2 (pp. 19-24) of the Report of the Ian Ramsey Centre on environmental ethics, *Values, Conflict and the Environment*, edited by Robin Attfield and Katharine Dell, Oxford: Ian Ramsey Centre and Centre for Applied Ethics, Cardiff, 1989. A survey of the range of principles of value currently adhered to in environmental matters.
66. ‘Reasoning About the Environment’, jointly composed with three members (Professor R.M. Hare, Dr. J.P. Griffin and Mr. A. Darby) of the Ian Ramsey Centre, St. Cross College, Oxford Working Party, and substantially revised by myself to meet criticisms from the other members; the main philosophical section (Chapter 3; pp. 25-46) of the Report of the Ian Ramsey Centre on environmental ethics, *Values. Conflict and the Environment*, edited by Robin Attfield and Katharine Dell, Oxford: Ian Ramsey Centre and Centre for Applied Ethics, Cardiff, 1989. In the light of the value-theory here defended, and of our contention that environmental values are commensurable, an enlarged and modified method of cost-benefit analysis is developed and defended with a view to implementation by planners and policy-makers.
67. ‘An Editorial Response’, Chapter 6 (pp. 83-86) of the Report of the Ian Ramsey Centre on environmental ethics, *Values, Conflict and the Environment*, edited by Robin Attfield and Katharine Dell, Oxford: Ian Ramsey Centre and Centre for Applied Ethics, Cardiff, 1989. In view of the criticisms expressed by two members of the Working Party of the main report in Chapter 5, the meaning and implications of earlier chapters are further clarified and defended.
68. ‘Secular Humanism’, paper commissioned for the Ian Ramsey Centre, St. Cross College, Oxford Working Party Report on environmental ethics, *Values, Conflict and the Environment*, edited by Robin Attfield and Katharine Dell, Oxford: Ian Ramsey Centre and Centre for Applied Ethics, Cardiff, 1989, of which it forms Chapter 8, pp. 91-96. The various distinctive historical stances and values of Enlightenment and post-Enlightenment humanism relating to future generations and to nonhuman nature are reviewed so as to exhibit both continuities and discontinuities with Judaic and Christian values.

69. 'Environmentalism', paper commissioned for the Ian Ramsey Centre, St. Cross College, Oxford Working Party Report on environmental ethics, *Values, Conflict and the Environment*, edited by Robin Attfield and Katharine Dell, Oxford: Ian Ramsey Centre and Centre for Applied Ethics, Cardiff, 1989, of which it forms Chapter 10, pp. 105-108. The diverse principles and values of different kinds of environmentalism are reviewed and compared.
70. 'Derek Parfit and the History of Ethics', in *History of the Human Sciences*, 2.4, October 1989, 357-371. Parfit's exposition of Christian ethics as founded on the principle of self-interest and his claims that till recently normative and applied ethics have been entrammelled by religion are contested, and alternative accounts both of Christian ethics and of recent developments in normative and applied ethics are offered and defended.
71. 'Deep Ecology and Intrinsic Value', *Cogito*, 4/1, 1990, 61-66; a response to a critic, which also explains the indispensability of the notion of intrinsic value.
72. 'The Global Distribution of Health Care Resources' *Journal of Medical Ethics*, 16/2, September 1990, 153-156; also in Harry Lesser (ed.), *Ageing, Autonomy and Ethics, Routledge*, 1999 (see 118, below). The international aspects of health care and developmental problems are investigated in the light of rival ethical paradigms, which are shown to produce convergent policy directives.
73. 'Etika ekologicheskoi otvetctvennosti' ('Ethics of environmental responsibility') (chapters from a book), trans. L.I. Vasilenko and V.Y. Koslov, in *Global Problems and Human Values*, Moscow: Progress Publishers, 1990, 203-257. A translation of chapters 2, 5 & 10 of Robin Attfield, *The Ethics of Environmental Concern* (see list of books).
74. 'Making Decisions', *Philosophy Now*, 1, 1991, 5 8. A discussion of methods, problems and principles relating to decision-making.
75. 'Has the History of Philosophy Ruined the Environment?', *Environmental Ethics*, 13.2, 1991, 127-137. Eugene Hargrove's critique of the impact of the history of ideas and the history of philosophy on attitudes to the environment is shown to be partially valid but partially in need of revision.
76. 'Attitudes to Wildlife in the History of Ideas', *Environmental History Review*, 15.2, 1991, 1-8. Eugene Hargrove's account of the history of attitudes to wildlife is appraised and subjected to qualification.

77. 'Collegiality and Efficiency in Universities', in John K. Roth and Creighton Peden (eds.), *Rights, Justice and Community*, Lewiston, NY: Edwin Mellen Press, 1992, selected papers from the Oxford 1988 Conference of the North American Society for Social Philosophy. The concept of efficiency employed by the Report of the Jarratt Commission is critically analysed, and an alternative model of university self-government commended for its capacity to uphold academic values.
78. 'Development and Environmentalism', in Barry Wilkins and Robin Attfield (eds.) *International Justice and the Third World: Essays in the Philosophy of Development*, London and New York: Routledge, 1992, 151-168, ISBN, 0-415-06924-6 (hb) & 0-415-06925-4 (pb). It is argued that morality and consistency oblige environmentalists to support sustainable development in the Third World, and that consistent developmentalists are obliged to support some of the deeper kinds of environmentalism.
79. 'Claims, Interests and Environmental Concern', in C.C.W. Taylor (ed.), *Ethics and the Environment*, Proceedings of Conference held at Corpus Christi College, Oxford, 20-21 September 1991, Corpus Christi College, Oxford, 1992; ISBN 0-9512844-l-X. In reply to Professor Bernard Williams it is argued that the interests of nonhuman animals must figure significantly both in environmental concern and in any satisfactory ethical system.
80. (Jointly with Barry Wilkins) 'Introduction', in Robin Attfield and Barry Wilkins (eds.) *International Justice and the Third World. Studies in the Philosophy of Development*, London and New York: Routledge, 1992, 1-16. The eight contributions are introduced.
81. 'Clarke, Independence and Necessity' *British Journal for the History of Philosophy*, 1.2, 1993, 67-82. While the version of the cosmological argument advanced in Clarke's Boyle Lectures is vulnerable to criticisms such as Hume's, the version of Clarke's critic Daniel Waterland remains defensible.
82. 'Sylvan, Fox and Deep Ecology: A View from the Continental Shelf', *Environmental Values* 2.1, 21-32, 1993. A critique of the philosophical literature concerning deep ecology and also of one of its foremost critics.
83. 'Development and Environmentalism' (see 78 above), in H. Odera Oruka (ed.), *Philosophy, Humanity and Ecology: Volume I. Philosophy of Nature and Environmental Ethics*, Nairobi: African Centre for Technology Studies Press, 1994, 133-149. ISBN 9966-41-086-4.

84. 'Rehabilitating Nature and Making Nature Habitable', in Robin Attfield and Andrew Belsey (eds.), *Philosophy and the Natural Environment*, Cambridge: Cambridge University Press, 1994, 45-57; ISBN 0-521-46903-1; ISSN (as special number of the journal Philosophy) 0031-8191(1994)69+1;1-P. While it is not impossible to restore nature, and both preservation and restoration are proper roles for humanity, they are subordinate to making nature sustainably habitable.
85. (Jointly with Barry Wilkins) 'Sustainability', *Environmental Values*, 3, 1994, 155-158. Arguments are adduced against understanding 'sustainability' as implying 'fit to be sustained'.
86. 'The Precautionary Principle and Moral Values', in Timothy O'Riordan and James Cameron (eds.), *Interpreting the Precautionary Principle*, London: Cameron & May, 1994, 152-164; ISBN 1-874698-45-7. A survey of the ethics of the precautionary principle.
87. 'Ethics and the Environment: the Global Perspective', in Brenda Almond (ed.), *An Introduction to Applied Ethics*, Basil Blackwell Publisher, 1994, 331-342, ISBN 0-631-19391-X. A critical review of the environmental ethics field.
88. (Jointly with Andrew Belsey) 'Introduction', in Robin Attfield and Andrew Belsey (eds.), *Philosophy and the Natural Environment*, Cambridge: Cambridge University Press, 1994, 1-12, ISBN 0-521-46903-1; ISSN 0031-8191(1994)69+1; 1-P. The sixteen contributions are introduced.
89. 'John Laird and the Idea of Value', *Journal of Value Inquiry*, 29, 1995, 103-114; ISSN 0022-5363. A commissioned re-evaluation of a classical text in value-theory.
90. 'Population Growth and Hope for Humanity', in Creighton Peden and Yeager Hudson (eds.), *The Social Power of Ideas*, Lewiston, NY, Queenston, Ontario and Lampeter, Dyfed: Edwin Mellen Press, 1995, 21-33. ISBN 0-7734-9043-4. A contribution to social theory, first made at the Tenth International Social Philosophy Conference, Helsinki, 1993.
91. 'Preferences, Health, Interests and Value', *Justifying Value in Nature, Special Topic Issue of The Electronic Journal of Analytic Philosophy*, 3:2, May 1995, 7-15 (of hard copy); ISSN 1071-5800. The capacity for health is argued to be sufficient for having interests and moral standing; thus the capacity for preferences is sufficient but unnecessary in these regards.

92. 'Development and Environmentalism', in Azizan Haji Baharuddin (ed.), *Environment and Development: Ethical and Educational Considerations*, Kuala Lumpur: Institute for Policy Research (IKD), 1995, 77-103. ISBN 983-884-035-1. (See 78, above.)
93. 'The Meaning of Names and Their Propositional Context', *Cogito*, 9.2, 1995, 153-157; ISSN 0950-8864. Michael Durrant has not shown the Context Principle of Frege and Wittgenstein to be false, but it should still be rejected.
94. 'Reasoning About the Environment: Ethical Decision-Making', in Azizan Haji Baharuddin (ed.), *Environment and Development: Ethical and Educational Considerations*, Kuala Lumpur: Institute for Policy Research (IKD), 1995, 177-198. ISBN 983-884-035-1. The method of Comprehensive Weighing is related to Third World contexts.
95. 'Genetic Engineering: Can Unnatural Kinds Be Wronged?', in Wheale, P.R. and McNally, R.M. (eds.) *Animal Genetic Engineering: Of Pigs, Oncomice and Men*, London: Pluto Press, 1995, 201-208. ISBN 074530754X (hb) and 0745307558 (pb). It is argued that transgenic animals can be harmed, and that also the ethics of generating them is partially independent of this possibility.
96. (Jointly with Susanne Gibson) 'Ethics', in Michael Payne (ed.), *Dictionary of Cultural and Critical Theory*, Oxford: Blackwell, 1996, 178-182; ISBN 0631-171975 (hb). Summarises the history and current issues of ethics (meta-ethical, normative and applied).
97. *Ethics and Problems for Sustainable Development*, Working Paper 77 of the Man and Nature Humanities Center, University of Odense, 1996, 15pp., ISSN 0907-1385; ISBN 87-89844-63-7. Strong Sustainable Development is defended against environmentalist and other objections.
98. 'Too High a Theme? Of Finitude, Predication and Analogy', *Scottish Journal of Religious Studies*, 17.1, 1996, 5-19. Language about God, like language used of nonhuman animals, can bear a sense analogous to that of the same language when used of human beings, despite the fears of the seventeenth-century poet Richard Baxter that, without supernatural help, the divine nature may be "too high a theme".
99. 'Climate Change and Ethics', *Philosophy Today*, No.22, May 1996, 8-9. Report of the Workshop of this title of the Society for Applied Philosophy, London, 3 February 1996.
100. 'El ambito de la moralidad' (an abridged version of chapter one of Robin Attfield, *A Theory of Value and Obligation*), trans. Carmen Velayos Castelo, in Jose Maria Garcia Gomez-Heras (ed.), *Etica del*

Medio Ambiente: Problema, Perspectivas, Historia, Madrid: Tecnos, 1997, 71-88; ISBN 84-309-2977-0. A study of the scope of moral standing.

101. 'Western Traditions and Environmental Ethics' (see 40, above), trans. Juhani Pietarinen, in Markku Oksanen and Marjo Rauhala-Hayes (eds.) *Ymparistofilosofia. Kirjoituksia ymparistonsuojun eettisista perusteista*, Helsinki: Gaudeamus-kirja, 1997, 47-72; ISBN 951-662-694-7.
102. 'Biozentrismus, moralischer Status und moralische Signifikanz' (trans. Dieter Birnbacher) (first published in 1986 as 'Biocentrism, Moral Standing and Moral Significance'), in Dieter Birnbacher (ed.), Okophilosophie, Stuttgart: Reclam, 1997, 117-133; ISBN 3-15-009636-7. (See 58, above.)
103. 'Ecophilosophy, Poverty and the Future', in Gilbert Ogutu, Pentti Malaska and Johanna Kojola (eds.), *Futures Beyond Poverty: Ways and Means Out of the Current Stalemate*, selected papers from the XIV World Conference of the World Futures Studies Federation (Nairobi, 1995), Helsinki: World Futures Studies Federation, 1997, 21-25; ISBN 951-738-747-4. Environmental philosophy need not be open to charges of indifference to poverty; its concern for future people commits its adherents to recognition of the needs of current people also and thus the requirements of social development.
104. 'Discounting, Jamieson's Trilemma and Representing the Future', in T. Hayward and J. O'Neill (eds.), *Justice, Property and the Environment: Social and Legal Perspectives*, Aldershot: Ashgate, 1997, 85-96; ISBN 1-85972-5295. Discounting is only justified where special reasons exist; relatedly, future people should be represented by proxies in decision-making.
105. 'Natur erhalten oder Menschen ernähren?', German translation of 'Saving Nature Versus Feeding People?' (not previously published), in *Conceptus: Zeitschrift für Philosophie* (Salzburg), XXIX, nr. 74, 1996 (published 1997), pp. 27-45; ISSN: 0010-5155. A criticism of Holmes Rolston's view that nature preservation should sometimes be given priority over saving human life.
106. 'Intrinsic Value and Transgenic Animals', in Andrew Johnson and Alan Holland (eds.), *Animal Biotechnology and Ethics*, a collection on the ethics of transgenic engineering, London: Chapman and Hall in 1998, 172-189; ISBN 0-412-75680-3 (hb), 0-412-59190-1 (pb). Coherent objections to transgenic engineering in terms of intrinsic value are consequentialist rather than Kantian.

107. 'Progress, Nature and Metaphysics', in Carsten Bengt-Pedersen and Niels Thomassen (eds.), *Nature and Lifeworld: Theoretical and Practical Metaphysics*, the proceedings volume of the Inter-Nordic Institute of Philosophy, Odense, 1995, Odense: Odense University Press, 1998, 11-26; ISBN: 87-7838-311-0. Some forms of belief in progress are compatible with recognition of nature's value, provided that nature is not construed as natural capital, and that its otherness is recognised.
108. 'Existence Value and Intrinsic Value', *Ecological Economics*, 24 (special number, edited by Bo Gustafsson of the Swedish Colloquium for Advanced Study in the Social Sciences, University of Uppsala), 1998, 163-8; ISSN: 0921-8009. Despite Jonathan Aldred's defence of existence value, such value fails to cover much of the extension of intrinsic value, and should be replaced in decision-making by direct consideration of items of intrinsic value, as represented by proxies.
109. 'Environmental Ethics (Overview)', in Ruth Chadwick (ed.), *Encyclopedia of Applied Ethics*, San Diego: Academic Press, 1998, vol. 2, 73-81. ISBN (of vol. 2): 0-12-227067-3. A historical and critical survey of the field of environmental ethics.
110. 'Saving Nature, Feeding People and Ethics', *Environmental Values*, 7, 1998, 255-68; ISSN: 0963-2719. Criticises Holmes Rolston's claim that nature preservation should sometimes be given priority over feeding people.
111. 'Environmental Ethics and Intergenerational Equity', *Inquiry*, 41, 1998, 207-22; ISSN: 0020-174X. Principles, policies and a model for relations between generations are discussed and sifted.
112. 'Responsibility for the Global Environment' (presented at Development Conference, Aberdeen, 1996), *International Journal of Applied Philosophy* (Florida) 12, 1998, 181-6; ISSN: 0739-098X. Argues that states have obligations to participate in treaties necessary to protect the global environment.
113. 'Gli atteggiamenti cristiani verso la natura' ('Christian Attitudes to Nature'), in Mariachiara Tallachini (ed.), *Etiche della terra: Antologia di filosofia dell' ambiente*, Milan: Vita e Pensiera, 1998, 103-127; ISBN 8834319028.
114. 'Aldo Leopold: A Brief Appraisal', *Reflections, Newsletter of the Program for Ethics, Science and the Environment* (Department of Philosophy, Oregon State University), 3, 1998, 7. Leopold is rightfully influential as an ecologist, but philosophically naive.

115. 'The Comprehensive Ecological Movement', in Edgar Morscher, Otto Neumaier and Peter Simons (eds.), *Applied Ethics: Issues and Foundations*, Dordrecht, Boston, London: Kluwer, 1998, 9-27. ISBN 0-7923-4965-2. A survey and critique of normative principles involving rejection of anthropocentrism, and of models for environmental decision-making.
116. 'Discounting, Jamieson's Trilemma and Representing the Future', (first published in 1997), republished in *Kriterion, Zeitschrift für Philosophie* (Salzburg, Austria), 12, (1997) 1999, 24-32; ISSN 1019-8288. (See 104, above.)
117. 'Humpty Dumpty, Carroll and Frege', *Cogito*, 13.1, 1999, 55-59; ISSN: 0950-8864. Compares Carroll's theory of stipulative meaning with Alice's common sense approach, in the light of Gottlob Frege's distinction between sense and reference.
118. 'The Global Distribution of Health Care Resources' in Harry Lesser (ed.), *Ageing, Autonomy and Ethics,* Routledge, 1999, 133-141; ISBN 1-84014-971-X. (See 72, above).
119. 'Depth, Trusteeship, and Redistribution', in Klaus Brinkmann (ed.), *The Proceedings of the Twentieth World Congress of Philosophy, Volume 1, Ethics*, Bowling Green, OH: Philosophy Documentation Center, 1999, 159-168; ISBN: 1-889680-05-2. Trusteeship is defended and shown to involve global restructuring.
120. 'Christianity', in Dale Jamieson (ed.), *A Companion to Environmental Philosophy*, Oxford and Malden, MA: Blackwell, 2000, 96-110; ISBN 1-55786-910-3. An historical overview of Christian attitudes to nature.
121. 'Rehabilitating Nature and Making Nature Habitable' (first published 1994), republished in William Throop (ed.), *Environmental Restoration: Ethics, Theory and Practice*; Amherst, New York: Humanity Books, 2000, pp. 113-126; ISBN 1-57392-818-6. (A republication of 84, above.)
122. 'Evolution, Theodicy and Value' (presented at the Science and Religion Forum Conference, Hoddesdon, 1996), *The Heythrop Journal*, 41.3, July 2000, 281-296; ISSN 0018-1196. Neither nature nor natural creatures can be regarded as immoral; nature, including predation and parasitism, is argued to be compatible with creation by a benign creator.
123. 'Society and Ethics: A Dialogue', *Philosophical Writings*, no. 13, Spring 2000, 19-28; ISSN 1361-9365. A dialogue between a moral philosopher and a sociologist.

124. 'The Good of Trees', republished in Peter C. List (ed.), *Environmental Ethics and Forestry: A Reader*, Philadelphia: Temple University Press, 2000, 98-113; ISBN 1-56639-784-7 (hb) and 1-56639-785-5 (pb); a book in the series 'Environmental Ethics, Values and Policy', ed. Holmes Rolston III.
125. 'Are Promises to Repay International Debt Binding?' (presented to International Society for Value Inquiry at World Congress of Philosophy, Boston, 1998), *Journal of Social Philosophy*, 32.4, Winter 2001, 505-511; ISSN 0047-2786. John Searle's findings about debt obligations do not imply that Third World countries have overriding obligations to service or repay debts.
126. 'Meaningful Work and Full Employment', *Reason in Practice*, 1.1, 2001, 41-48; ISSN, 1473-589X. Arguments from essential capacities and from self-respect continue to support the value of meaningful work, for which full employment is indispensable.
127. 'Sustainability, Global Warming, Population Policies and Liberal Democracy', in Marcel Wissenburg and John Barry (eds.), *Sustaining Liberal Democracy: Ecological Challenges and Opportunities*, Basingstoke and New York: Palgrave, 2001, 149-160. ISBN 0-333-91981-5. Liberal democracy is argued to be compatible with sustainable policies for greenhouse gas emissions and for population.
128. 'Postmodernism, Value and Objectivity', *Environmental Values*, 10.2, 2001, 145-162; ISSN 0963-2719. Belief in objective intrinsic value is defended against postmodernist and relativist critiques.
129. 'The Global Distribution of Health Care Resources' (72 above), in Ruth Chadwick and Doris Schroeder (eds), *Applied Ethics* (vol.3), London, Routledge, 2001, 231-238; ISBN 0-415-20837-8.
130. 'To Do No Harm: The Precautionary Principle and Moral Values' (a revision of 86 above), *Reason and Practice*, 1.3, 2001, 11-20; ISSN, 1473-589X.
131. 'The Good of Trees', in John O'Neill, R. Kerry Turner and Ian J. Bateman (eds.), *Environmental Ethics and Philosophy*, Cheltenham: Edward Elgar, 2001, 328-347, ISBN 1-84064-221-1. (A republication of 34 above.)
132. 'Global Citizenship and the Global Environment', in Nigel Dower and John Williams (eds.), *Global Citizenship*, Edinburgh: Edinburgh University Press, 2002, 191-200; ISBN, 0-7486-1547-4. Global citizens recognise universal obligations, not least with regard to the shared, objective global environment.

133. 'Ecological Policies and Ecological Values', trans. K. Boudouris as 'Oikologikes politikes kai oikologikes axies' in *Oikologikes Axies*, ed. Konstantine Boudouris, Athens: University of Athens Technology Institute of Applied and Ecological Philosophy, University of Athens, 2002, 9-26; ISBN, 960-918440-5. New grounds are offered for preferring biocentrism as a value-theory to anthropocentrism.
134. 'The Good of Trees', in David Schmidtz and Elizabeth Willott (eds.), *Environmental Ethics: What Really Matters, What Really Works*, New York and Oxford: Oxford University Press, 2002, pp. 58-71; ISBN 0-19-513909-7. (A republication of 34 above.)
135. 'Saving Nature, Feeding People and Ethics', in Andrew Light and Holmes Rolston III (eds.), *Environmental Ethics: An Anthology*, Cambridge, MA: Blackwell, 2002, 463-471; ISBN 0-631-22294-4. (A republication of 110 above.)
136. 'Ecological Sustainability in a Developing Country such as South Africa? A Philosophical and Ethical Inquiry' (jointly authored with Johan Hattingh), *The International Journal of Human Rights*, 6.2 (Summer 2002), pp. 65-92; ISSN 1364-2987. Conceptual and ethical problems for ecologically sustainable development are shown not to prevent such development being reconcilable with social justice, even in a developing country such as South Africa.
137. 'Verification, God-Talk and Metaphysics', *The New Theologian*, 13.1, Winter 2002, 4-5. The Verification Principle, far from undermining metaphysics, undermines itself; there is nothing incoherent in reasoning about what transcends verification.
138. 'Global Warming, Justice and Future Generations', *Philosophy of Management*, 3.1, 2003, 17-23; ISSN, 1473-589X. Principles for carbon emission regimes are sifted, and an egalitarian, population-based principle commended.
139. 'Biocentric Consequentialism, Pluralism and 'the Minimax Implication': A Reply to Alan Carter', *Utilitas*, 15.1 (March 2003), 76-91; ISSN 0953-8208. Biocentric consequentialism is defended against criticisms from Alan Carter relating to population, quality of life and environmental sensitivity.
140. 'Global Warming, Justice and Future Generations', in Harto Hakovirta (ed.), *Six Essays on Global Order and Governance*, Turku: Academy of Finland Figare/Safir Project, Department of Political Science, University of Turku, 2003, 71-86; ISBN 951-29-2600-8. A republication of 138 (above).

141. 'Environmental Problems and the Ethics of Science', in Lebanese National Commission for UNESCO (ed.), *Ethics of Science and Technology*, Beirut: Lebanese National Commission for UNESCO, 2003, 150-156; ISBN 9953-0-0209-6. A discussion of ethical aspects of regional problems including fresh water, energy and global warming.
142. 'Differentiated Responsibilities', in Markku Oksanen and Juhani Pietarinen (eds.), *Philosophy and Biodiversity*, Cambridge: Cambridge University Press, 2004, 237-250; ISBN: 0-521-80430-2 and (pbk), 2007, (978)0521039147. Responsibilities are both universal and geared to the situations of agents, as the Rio Declaration (1992) recognises.
143. 'Sustainable Development, Sustainable Livelihoods and Land Reform in South Africa: A Conceptual and Ethical Inquiry' (jointly authored with Johan Hattingh and Manamela Matshabaphala), in *Third World Quarterly*, 25.2, 2004, 405-21; ISSN 0143-6597. The relations between sustainable development, sustainable livelihoods and land reform are explored, and the case against and that in favour of land reform in South Africa are presented and appraised. A 'research highlight' in which Rabi Thapa re-presents and highlights this article has been published (18/5/05) on id21, the DfID website, at http://www.id21.org/society/s1ara1.html; a further abstract has also appeared in the Sociological Abstracts database of CSA Illumina (June 2005).
144. 'The Concept of Sustainable Development Revisited', in *Yeditepe 'de Felsefe* (Philosophy at Yeditepe University, Istanbul), 1.3, 2004, 300-309; ISSN 1304-0197. The radical core of this concept is defended against both economistic interpretations and charges of irremediable vagueness.
145. 'Rousseau, Clarke, Butler and Deism', *British Journal for the History of Philosophy*, 12.3, 2004, 429-443; ISSN 0960-8788. Rousseau is argued to have produced a defence of deism sufficient to uphold it against eighteenth-century criticism and make it still a potential option amongst metaphysical positions.
146. 'The Good of Trees', in J. Baird Callicott and Clare Palmer (eds), *Environmental Philosophy: Critical Concepts in the Environment* (5 vols), London: Routledge, 2004, vol. 1, ISBN: 0-415-32646-X. A republication of 34 above.
147. 'Nationalism, Global Citizenship, Environmental Values, and the Common Heritage of Humanity', in Jouni Paavola and Ian Lowe (eds),

Reconceiving Environmental Values in a Globalizing World, London: Routledge, 2005, 38-50; ISBN 0415343623. (Paperback, 2007, ISBN: 9780415459877; ISBN10: 0415459877.) Defends a cosmopolitan but non-anthropocentric ethic and attitudes of global stewardship as historically precedented and urgently required for coping with shared global problems.

148. 'J.R. Lucas', *Dictionary of Twentieth-Century British Philosophers* (2 vols.), ed. Stuart Brown, Bristol: Thoemmes Continuum, 2005, vol. I, 586-90; ISBN 1-84371-096-X. A 2500-word account and appraisal of the work of this major philosopher.

149. 'Environmental Ethics, Environmental Problems and the Ethics of Science', in *Ethics, Law and Society*, Volume 1 (eds. Jennifer Gunning and Søren Holm) Aldershot: Ashgate, August 2005, 95-101; ISBN: 0-7546-4583-5. A revision of 141 above.

150. 'Biocentric Consequentialism and Value-Pluralism: A Response to Alan Carter', *Utilitas*, 17.1, March 2005, 85-92; ISSN: 0953-8208. A rejoinder to Alan Carter's criticisms of biocentric consequentialism and advocacy of value-pluralism.

151. 'Future Generations: Considering All the Affected Parties', translated into Spanish by Adrián Pradier, under the supervision of Carmen Velayos, as 'Generaciones futuras: considerando todes das partes afectadas', in *Isegoria* (Spain), 32, junio 2005 (published December 2005), 35-46; ISSN, 1130-2097. Current agents have responsibility for differences that can be made to the quality of life of future people and other future creatures, insofar as these are foreseeable.

152. 'Sustainable Forests, Global Responsibility and the Earth Charter', *Silva Carelica* (Finland), 49, 2005, 11-25; ISSN 0780-8232. Sustainable development must be global, and based on an ethic such as that of the Earth Charter, recommended here to foresters.

153. 'In Defense of *Environmental Ethics*', *Environmental Ethics*, 27, 2005, 335-36; ISSN, 0163-4275. My book *Environmental Ethics* is defended against a reviewer whose describes it unrecognisably.

154. 'Unprojected Value, Unfathomed Caves and Unspent Nature', *Environmental Values*, 14.4, 2005, 513-18; ISSN, 0963-2719. Passages of poetry are used to illustrate that not all value is to be regarded as a projection of human sentiments.

155. 'Altruism and Environmental Concern' (joint paper with Stephen Moller), *Copula, Journal of the Department of Philosophy, Jahangirnagar University* (Bangladesh), 22, June 2005, 51-59; ISSN,

1563-0692. Diverse theories of environmental ethics are tested for their coherence with the phenomenon of altruism.

156. 'Environmental Sensitivity and Critiques of Stewardship', in R.J. Berry (ed.), *Environmental Stewardship: Critical Perspectives—Past and Present*, London: T&T Clark/Continuum, January 2006, 76-91; ISBN 0567031172 (hb), 0567030180 (pb). Religious and secular versions of belief in human stewardship of the earth are clarified and defended against several varieties of criticism.
157. 'Environmental Ethics and Global Sustainability', in Henk A.M.J. ten Have (ed.), *Environmental Ethics and International Policy*, Paris: UNESCO, 2006, 69-87, ISBN-10: 92-3-104039-1; ISBN-13: 978-92-3-104039-9 (in a UNESCO series on Ethics of Science and Technology; a volume arising from UNESCO Expert Working Party on environmental ethics meetings in Paris and New Orleans of 2004). An analysis of the state of the art of environmental ethics and of implications for international action.
158. 'Development and Environmentalism', in Christine Koggel (ed.), *Moral Issues in Global Perspective* (2nd edn, 3 vols.), vol. 3 (*Moral Issues*), Guelph (Canada): Broadview Press, 2006, 272-282; ISBN 1551117495. A republication of 78 above.
159. 'The Shape of a Global Ethic', *Philosophy & Social Criticism*, 32.1, 2006, 5-19; ISSN: 0191-4537; DOI (Digital Object Identifier): http://dx.doi.org/10.1177/0191453706059843. Problems including future-related responsibilities preclude most kinds of ethical theory from suitability as a global ethic, but practice-consequentialism is argued to overcome such problems, provided it adopts a biocentric form.
160. 'Is the Concept of Nature Dispensable', *Ludus Vitalis: Journal of Philosophy of Life Sciences* (Spain), vol. xiv, no. 25, June 2006, 105-1167; ISSN: 1133-5165. Also to be translated into Spanish, for a collection to be edited by Teresa Kwiatkowska. The concept of nature is argued to be indispensable, not least for ethics, despite the claims of strong constructivists.
161. 'Applied Ethics', in A.C. Grayling, Andrew Pyle and Naomi Goulder (eds), *Continuum Encyclopedia of British Philosophy* (4 vols), New York: Thoemmes Continuum, 2006, vol. I, 124-129; ISBN: 1-84371-141-9. An overview of applied ethics in Britain from the seventeenth century.
162. 'Leibniz, the Cause of Gravity, and 'Physical Theology'', in *Studia Leibnitiana* 36.2, 2005 (published 2007), 238-244; ISSN, 0039-3185.

Farrer's category of physical theology is applied to Leibniz's Newtonian opponents.

163. 'Is the Concept of Nature Dispensable?', in Stephen Voss, Berna Kylync and Gurol Irzyk (eds), *Logic and Philosophy of* Science, vol. 5 of *Proceedings of the XXI World Congress of Philosophy*, based on the 2003 World Congress of Philosophy (Istanbul), Ankara: Philosophical Society of Turkey, 2007, 59-63; ISBN 978-975-7748-34-2 (TK No); 978-975-7748-42-7 (5.C). (A shorter version of 160 (above).)

164. 'Sustainable Development Revisited', in Zeynep Davran (ed.), *Human Rights*, Volume 3 of *Proceedings of the XXI World Congress of Philosophy*, based on the 2003 World Congress of Philosophy (Istanbul), Ankara: Philosophical Society of Turkey, 2007, 185-189; ISBN 978-975-7748-34-2 (TK No); 978-975-7748-48-9 (3.C). (A shorter version of 144 (above).)

165. 'Cudworth, Prior and Passmore on the Autonomy of Ethics', forthcoming in Sarah Hutton and Douglas Hedley (eds), *Platonism at the Origins of Modernity*, published in the series International Archives in the History of Ideas, Dordrecht: Springer, 2008; ISBN 978-1-4020-6407-7 (hb); 978-1-4020-6407-4 (e-book).

166. 'Lynn White, Jr, 1907-1987', in J. Baird Callicott and Robert Frodeman (eds), *Encyclopedia of Environmental Ethics and Philosophy*, Farmington Hills, MI: Thomson Gale, 2008, vol. II, 400-402. Conveys central ideas and criticisms of Lynn White and his well-known thesis.

167. 'Hunger', in J. Baird Callicott and Robert Frodeman (eds), *Encyclopedia of Environmental Ethics and Philosophy*, Farmington Hills, MI: Thomson Gale, 2008, vol. I, 495-499. Discusses the nature and causes of world hunger and related obligations in an environmental context.

168. 'Biocentrism', in J. Baird Callicott and Robert Frodeman (eds), *Encyclopedia of Environmental Ethics and Philosophy*, Farmington Hills, MI: Thomson Gale, 2008, vol. I, 97-100. Discusses the varieties of biocentric thought and objections thereto.

169. 'Beyond the Earth Charter: Taking Possible People Seriously', *Environmental Ethics*, 29.4, Winter 2007 (published February 2008), 359-367; ISSN, 0163-4275. The Earth Charter needs to be strengthened to cope with the problem that future obligations are not owed to identifiable individuals, but to whoever lives in future times.

170. 'Mediated Responsibilities, Global Warming and the Scope of Ethics', *Journal of Social Philosophy*, 40.2, 2009, 225-236; ISSN 0047-2786. Both individuals and governments are responsible even when the impacts of their action or inaction are mediated.
171. 'Non-Reciprocal Responsibilities and the Banquet of the Kingdom', *Journal of Global Ethics*, 5.1, 2009, 33-41; ISSN (print) 1744-9626; (online) 1744-9634. In reply to Chris Groves, the ethics-of-care model is argued to stand in need of supplementation with regard to non-reciprocal responsibilities from other models, including that of hospitality.
172. 'Social History, Religion and Technology: An Interdisciplinary Investigation into White's 'Roots'', *Environmental Ethics*, 31.1, 2009, 31-50; ISSN, 0163-4275. White's controversial 'Roots' paper involves, pace Harrison, a critique of Christianity that misses the mark, while, as Whitney argues, disregarding economic and institutional causes of ecological problems, yet relevantly focusing on the influence of values and beliefs.
173. 'Ecological Issues of Justice', *Journal of Global Ethics*, 5.2, 2009, 147-154; ISSN (print) 1744-9626; (online) 1744-9634. A study of the implications for global justice and global warming of the wide scope of the concepts of 'moral patient' and 'moral agent', which also compares the Greenhouse Development Rights approach with the approach of Contraction and Convergence.
174. 'Exploring Ethical and Religious Attitudes to Sustainability', in Deva Temple and Olivier Urbain (eds), *Ethical Transformations for a Sustainable Future* (the book version of *Peace & Policy*, 14, 2009), Piscataway, NJ: Transaction, 33-40; ISBN 978-1-4128-1445-4.
175. 'Philosophy on Poetry, Philosophy in Poetry', in Jinfen Yan and David Shrader (eds), *Creating a Global Dialogue on Value Inquiry: Papers from the XXII World Congress of Philosophy*, Lewiston, NY and Lampeter, UK: Edwin Mellen Press, 2009, 417-432; ISBN 978-0-7734-4702-8. A discussion of whether Lucretius' philosophical poem would be countenanced in Plato's *Republic*.
176. 'Mediated Responsibilities, Global Warming and the Scope of Ethics', in Ruth Irwin (ed.), *Climate Change and Philosophy*, London: Continuum, 2010, 183-216; ISBN 978-0-8264-4065-5. (An expanded version of 170 (above).)
177. 'The Precautionary Principle', Guest Editorial, *UK Centre for Bioscience Bulletin*, 29, Spring 2010, 1, ISSN 1740-6692 (print);

1740-6706 (online). All science students should be taught about the Precautionary Principle.

178. 'Sustainability', forthcoming in Hugh LaFollette (ed.), *International Encyclopedia of Ethics* in 2010. Presents a non-normative account of sustainability.

179. 'Biocentrism', forthcoming entry in *Encyclopedia of Environmental Issues.* An account of the common features and diverse versions of biocentrism in environmental philosophy.

180. 'Brundtland Commission', forthcoming entry in *Encyclopedia of Environmental Issues.* A short depiction of *Our Common Future*, produced by the World Commission on Environment and Development in 1987.

181. 'Darwin, Meaning and Value', forthcoming in *Environmental Values*, 19.4, November 2010. A reply to some criticisms from Alan Holland about the role of value in life being seen as meaningful.

182. 'Darwin's Doubt, Non-determinstic Darwinism and the Cognitive Science of Religion', forthcoming in *Philosophy.* Plantinga's argument that evolutionary theory makes our beliefs unreliable applies to deterministic versions of Darwinism and of the cognitive science of religion, but non-deterministic versions (like those of Rose, Lewontin, Ward and Miller) escape it.

183. 'The Global Distribution of Health Care Resources in the Twenty-First Century', to be published in a Rodopi collection to be edited by Harry Lesser.

184. 'Climate Change: The Ethical Dimension', forthcoming in *Ethics and Climate Change: Scenarios for Justice and Sustainability*, edited by Lucia Mariani and Matteo Mascia, a collection based on the Fondazione Lanza international conference on climate change, Padua, October 2008.

185. 'Climate Change, Environmental Ethics, and Biocentrism' forthcoming in Ved Nanda and Olivier Urbain (eds), *Facing Climate Change with a New Environmental Ethic*, Piscataway, NY: Transaction. A biocentric approach is defended and shown to justify significant action on climate change.

186. 'Social History, Religion and Technology: An Interdisciplinary Investigation into White's 'Roots'' (172 above), to be republished in a collection on anthropocentrism to be edited by Rob Boddice.

187. 'Ecological Issues of Justice' (see 173 above), forthcoming in Heather Widdows and Nicola Smith (eds), *Global Social Justice*, to be published by Routledge.

188. 'Has the History of Philosophy Ruined the Environment?' (75 above), to be republished online by Thomas Seiler at www.umweltethik.at
189. 'Reconciling Individualist and Deeper Environmentalist Theories? An Exploration', forthcoming in Patrick Blandin and Donato Bergandi (eds), *Ecology, Evolution, Ethics: The Virtuous Epistemic Circle*, Paris: National Museum of Natural History. A critique of Alan Carter's pluralism.
190. 'Global Warming, Equity and Future Generations', forthcoming in *Human Ecology Review*. A defence of equal per capita carbon emissions entitlements and of Contraction and Convergence, as presented at the 22nd World Congress of Philosophy (Seoul, Korea, 2008).
191. 'Intrinsic Value and Transgenic Animals', forthcoming in *Being for the Other: Issues in Animal Rights and Ethics*, Manish A. Vyas (ed.), to be published by Daya Publishing House, New Delhi. A republication of 106 (above).

Reviews

1. Review of D.D. Raphael (ed.), *British Moralists 1650-1800*, Oxford: Clarendon Press, 1969, in *The Church Quarterly*, April 1970, p.363 (book notes section: unsigned).
2. Review of Robert A. Evans (ed.), The *Future of Philosophical Theology*, Philadelphia: The Westminster Press, 1971, in *Religious Studies*, 9.1, March 1973, 99-101.
3. Review of William A. Christian, *Oppositions of Religious Doctrines*, Macmillan, 1972, in *Trivium*, 8, 1973, 145-47.
4. Further review of *Oppositions of Religious Doctrines* (see 3 above), in *Religious Studies*, 10.1, March 1974, 122-24.
5. Review of J. O'Higgins (ed.), *Determinism and Freewill*, Anthony Collins, *A Philosophical Inquiry Concerning Human Liberty*, The Hague: Martinus Nijhoff, 1976, in *Studia Leibnitiana*, 12.2, 1980, 293-96.
6. Review of Joseph Margolis, *Persons and Minds: The Prospects of Nonreductive Materialism*, Boston Studies in the Philosophy of Science, Volume LVII, Dordrecht, Holland; Boston, USA; London, England: Reidel, 1978, in *Studia Leibnitiana*, 12.2, 1980, 293-96.

7. Review of Renford Bambrough, *Moral Scepticism and Moral Knowledge*, London and Henley: Routledge & Kegan Paul, 1979, in *Philosophical Quarterly*, 31, 1981, 177-78.
8. Review of Steven T. Davis (ed.), *Encountering Evil. Live Options in Theodicv*, Edinburgh: T. & T. Clark, 1981, in *Insight, A Journal for Church and Community*, 1, 1982, 38-40.
9. Review of Keith Ward, *Holding Fast to God: A Reply to Don Cupitt*, London, SPCK, 1982, in *New Blackfriars*, 64, No. 755, 1983, 237-39.
10. Review of Stephen T. Davis, *Logic and the Nature of God*, London and Basingstoke: The Macmillan Press, 1983, in *New Blackfriars*, 65, No. 764, 1984, 90-92.
11. Review of R.G. Frey, *Rights, Killing and Suffering*, Oxford: Basil Blackwell Publisher, 1983, in *The Friend*, 142.11, March 16, 1984, 335-36.
12. Review of Nigel Dower, *World Poverty: Challenge and Response*, York: Ebor Press, 1983, in *Journal of Applied Philosophy* 1.2, 1984-5, 322-23; ISSN 0264-3758.
13. Review of Keith Thomas, *Man and the Natural World*, New York: Pantheon Books (and London: Allen Lane), 1983, in *Isis* (Official Journal of the History of Science Society), 75:3:278, 1984, 588-89.
14. Review of Grace M. Jantzen, *God's World, God's Body*, London: Darton, Longman and Todd, 1984, in *New Blackfriars*, 65, No. 771, 1984, 395-96.
15. Review of William Tucker, *Progress and Privilege. America in the Age of Environmentalism*, Garden City, New York: Anchor Press/Doubleday, 1982, in *Environmental Ethics*, 7.2, 1985, 181-83.
16. Review of Keith W. Clements, *A Patriotism for Today: Dialogue with Dietrich Bonhoeffer*, Bristol Baptist College, 1984, in *The Modern Churchman*, 27.1, 1985, 67-68.
17. Review of J. Kellenberger, *The Cognitivity of Religion: Three Perspectives*, Basingstoke and London; Macmillan, 1985, in *New Blackfriars*, 67, No. 792, 1986, 296-98.
18. Review of Roger Trigg, *Understanding Social Science. A Philosophical Introduction to the Social Sciences*, Oxford and New York: Basil Blackwell, 1985, in *Philosophy*, 61, 1986, 544-46.
19. Review of Mary Midgley, *Evolution as a Religion: Strange Hopes and Stranger Fears*, London and New York: Methuen, 1985, in *Philosophical Books*, 28, 1987, 118-20.

20. Review of Charles Whitney, *Francis Bacon and Modernity*, New Haven and London: Yale University Press, 1986, in *Journal of the History of Philosophy*, 26.4, October 1988, 665-67.
21. Review of Andrew Linzey, *Christianity and the Rights of Animals*, London, SPCK, 1987, in *Theology*, November 1988, 549-50.
22. Review of Jenny Teichman, *Pacifism and the Just War: A Study in Applied Philosophy*, Oxford: Basil Blackwell, 1986, in *Philosophical Books*, 29.2, April 1988, 103-105.
23. Review of William A. Christian, *Doctrines of Religious Communities: A Philosophical Study*, New Haven and London: Yale University Press, 1987, in *Philosophical Books*, 29.4, October 1988, 252-53.
24. Review of Andrew Brennan, *Thinking About Nature: An Investigation of Nature: Value and Ecology*, London: Routledge and Athens: University of Georgia Press, 1988, in *Journal of Applied Philosophy*, 6.2, 1989, 237-8.
25. Review of Holmes Rolston, *Environmental Ethics: Duties to and Values in the Natural World*, Philadelphia: Temple University Press, 1988, in *Environmental Ethics*, 11.4, Winter 1989, 363-68.
26. Review of Nigel Dower (ed.), *Ethics and Environmental Responsibility*, Aldershot: Avebury, 1989, in *Philosophical Books*, 31.3, 1990, 177-178.
27. Review of David Carr, *Educating the Virtues*, London and New York: Routledge, 1991, in *The Philosophical Quarterly*, 42, no. 168, 1992, 379-382.
28. Review of Ellen Frankel Paul, Fred D. Miller Jr and Jeffrey Paul (eds.), *The Just Society*, Cambridge, New York and Melbourne: Cambridge University Press, 1995, in *The Expository Times*, 108.3, December 1996, 91a-91b; ISSN 0014-5246.
29. Review of Holmes Rolston III, *Genes, Genesis and God*, Cambridge, New York and Melbourne: Cambridge University Press, 1999, in *Reason and Practice*, 1.1, January 2001, 75-77; ISSN 1473-589X.
30. Review of Holmes Rolston III, *Genes, Genesis and God*, Cambridge, New York and Melbourne: Cambridge University Press, 1999, in *Science and Religion Forum Reviews*, 37, July 2001, 29-31; ISSN 0967-5981. (A republication of 29 above.)
31. Review of Partha Dasgupta, Human Well-Being and the Natural Environment, Oxford and New York: Oxford University Press, 2001, in *Philosophy*, 78, 2003, 123-127; ISSN 0031-8191.

32. Review of Nigel Dower, *An Introduction to Global Citizenship*, Edinburgh: Edinburgh University Press, 2003, in *Philosophy of Management*, 5.3, 2005, 126-127; ISSN 1740-3812.
33. Review of John O'Neill, Alan Holland and Andrew Light, *Environmental Values*, London and New York: Routledge, 2008, ISBN 978-0-415-14509, in *Environmental Sciences*, 5.3, 2008, 1-2; ISSN 1569-3430 (print) / ISSN 1744-4225 (online).
34. Review of Michael Murray, *Nature, Red in Tooth and Claw: Theism and the Problem of Animal Suffering*, Oxford University Press, 2008, in *Reviews in Science and Religion*, 53, May 2009, 37-40; ISSN 1754-4742.
35. Review of James Maclaurin and Kim Sterelny, *What Is Biodiversity?* , Chicago and London: University of Chicago Press, 2008, in *Philosophy*, 84, 2009, 605-609; ISSN 0031-8191.
36. Review of Christopher J. Preston, *Saving Creation: Nature and Faith in the Life of Holmes Rolston III*, San Antonio: Trinity University Press, 2009, forthcoming in *Environmental Ethics*; ISSN (print) 0163-4275; (online) 2153-7895.

CONTRIBUTORS

Robin Attfield taught Philosophy at Cardiff University from 1968 to 2009, and became a Professor of Philosophy there in 1992. He has also taught in Nigeria, Kenya and Oregon. His contribution to environmental philosophy was recognised in 2008 with the award of a Cardiff University D.Litt. His publications are listed elsewhere in this volume. He has three children and seven grand-children. He and his wife Leela are members of the Religious Society of Friends (Quakers).

Roger Crisp is Uehiro Fellow and Tutor in Philosophy at St Anne's College, Oxford, and Professor of Moral Philosophy at the University of Oxford. He works mainly in ethics and the history of ethics. He is the author of *Mill on Utilitarianism* (Routledge, 1997) and *Reasons and the Good* (Clarendon Press, 2006). He has translated Aristotle's *Nicomachean Ethics* for Cambridge University Press (2000). He is the Philosophy Delegate to Oxford University Press, an Associate Editor of *Ethics*, and a member of the *Analysis* Committee.

Nigel Dower is Honorary Senior Lecturer in Philosophy at the University of Aberdeen, Scotland where he taught for most of the period 1967-2004. He has also been a visiting professor a number of times in America and Iceland. He now acts as an academic consultant on *Cosmopolitan agendas—ethics in a globalized world*. He was President of the International Development Ethics Association from 2002 to 2006. His research interests in the last twenty years have focussed on various issues in global ethics, including development, the environment, human rights, war and peace, and global citizenship. His publications include *World Ethics—the New Agenda* (1998; 2nd edition, 2007), *Introduction to Global Citizenship* (2003) and *The Ethics of War and Peace* (2009). In 2007 he received an Honorary Doctorate (TD) from the University of Uppsala for his work on global ethics and related issues. Nigel Dower is married, has three children and one grandchild, has been a Quaker since 1980, and his other interests include walking and gardening.

Alan Holland is Emeritus Professor of Applied Philosophy at Lancaster University. During the earlier part of his career he worked mostly on problems in epistemology and philosophical logic. Latterly his interest has turned to problems of a more practical or "applied" kind, such as those relating to the beginning of (human) life, to animals, and to the environment. He was founding editor of the journal *Environmental Values*, and recently co-authored a book on this topic with John O'Neill and Andrew Light. Current topics of research include: 'How to talk about the experiences of animals', 'The concept of the unnatural', and 'Why it is important to take account of history in environmental decision-making'. Some sample publications are:

1977 'Scepticism and causal theories of knowledge', *Mind* 86: 555-73.

1984 'On behalf of moderate speciesism', *Journal of Applied Philosophy* 1: 281-91.

1990 'A fortnight of my life is missing: reflections on the status of the human "pre-embryo"', *Journal of Applied Philosophy* 7: 25-37.

1994 'Natural capital', in *Philosophy and the Natural Environment* (eds. A. Belsey and R. Attfield) Cambridge University Press: 169-182.

1997 'Fortitude and tragedy: the prospects for a Stoic environmentalism', in *The Greeks & the Environment* (eds. T. Robinson and L. Westra) Rowman & Littlefield: 151-166.

2002 'Are choices trade-offs?', in *Economics, Ethics & Environmental Policy* (eds. D. Bromley & J. Paavola). Blackwell: Oxford. 17-34.

2009 'Darwin and the meaning in life', *Environmental Values* 18: 503-518.

Rebekah Humphreys worked at Cardiff University as a Tutor in Philosophy for a number of years. She completed a Master of Arts degree in Ethics and Social Philosophy at Cardiff University before completing a Ph.D. in Philosophy, specialising in animal ethics, at the same university. Her publications include 'Game Birds: the ethics of shooting birds for sport', in *Sport, Ethics and Philosophy*: *Journal of the British Philosophy of Sport Association* (published by Routledge: Taylor and Francis Group), Vol. 4, No.1, Apr. 2010, pp. 52-65. Her research interests include environmental ethics and animal ethics. She plays an active role in campaigns against animal abuse and animal exploitation.

Mary Midgley's special interests are in the relations between humans and the rest of nature, and in the troubled frontier between science and religion (particularly in cases where science becomes a religion). Her belief is that academic philosophy should be brought back into connexion with life, rather than becoming a form of highbrow chess for graduate students.

Among her books are *Beast and Man* (Routledge 1995), *Evolution as a Religion* (Methuen 1985), *Science as Salvation* (Routledge 1992) and *Science and Poetry* (Routledge 2001). Her most recent book is a memoir called *The Owl of Minerva* (Routledge 2005). At present she is working on the concept of Gaia.

Born in 1919, Mary Midgley took her university degree in Classics, Philosophy and Ancient History at Oxford, during the war. After graduate work, she lectured in Philosophy, first at the University of Reading, then (having married another philosopher, Geoffrey Midgley) at that of Newcastle on Tyne, where she became a Senior Lecturer. She still lives in Newcastle. She has three sons.

Besides miscellaneous broadcasting and book-reviewing, Mary has campaigned for the Anti-Nuclear Movement and for various causes on behalf of animals. For some years she chaired the RSPCA's advisory committee on Animal Experimentation and also the Forum for Science and Religion.

Clare Palmer is professor of Philosophy at Texas A&M University. She studied at Oxford University, both as an undergraduate at Trinity College, where she gained a First Class degree in Theology, and as a postgraduate at Wolfson and at Queen's College, where she was awarded a D.Phil. She has worked at Stirling University, Lancaster University, and most recently at Washington University in St Louis. Her publications include *Animal Ethics in Context* (Columbia University Press, 2010) *Animal Rights* in Ashgate's International Library of Essays on Rights (2008) and *Environmental Ethics and Process Thinking* (Oxford University Press, 1998).

Christopher Southgate trained originally as a biochemist at Cambridge, and has subsequently been a house-husband, a bookseller, and a lay chaplain. He is the author of five collections of poetry, including a verse biography of T.S. Eliot, and *Easing the Gravity Field: poems of science and love*. He is currently Lecturer in Theology at the University of Exeter, and Dean of Studies of the South West Ministry Training Course. Dr Southgate has edited a major textbook on science and religion, *God, Humanity and the Cosmos* (T&T Clark, 1999, 2005) and is the author of *The Groaning of Creation; God, evolution and the problem of evil* (Westminster John Knox Press, 2008). He is also the co-author of *Greening Paul: re-reading the Apostle in a time of environmental crisis* (Baylor U. Pr., 2010).

Sophie Vlacos recently completed a doctoral dissertation on the hermeneutic philosopher Paul Ricoeur and his relationship to literary theory. Her interests include questions of fictional reference, Modernist poetry, and the natural world.

INDEX